Aquaponics

Aquaponics

Farming Fish and Growing Greens for Grades 9–12

Amy M. Shambach,
Rebecca M. Ward, J. Stuart Carlton,
Kwamena K. Quagrainie, & Tomas O. Höök

Purdue University Press | West Lafayette, Indiana

Cataloging-in-Publication Data on file at the Library Congress.

978-1-62671-247-8 (paperback)
978-1-62671-248-5 (epdf)

Cover image: Rows of leafy green lettuce grow in blue hydroponic beds inside a greenhouse.
Courtesy of Illinois–Indiana Sea Grant.

Contents

The Sea Grant Aquaponics Curriculum

ILLINOIS–INDIANA

This curriculum was designed to be used in conjunction with an onsite aquaponics system. The lessons are aligned with Next Generation Science Standards, incorporating material from biology, chemistry, and environmental science. The general idea of the curriculum is for students to learn basic scientific concepts through the lens of an aquaponics system. In this way, standards are met while also demonstrating their relevance to an applied system as well as providing an opportunity for genuine experience and experimentation. While some lessons naturally favor a specific discipline, an effort was made to draw connections between the sciences to showcase the true interdisciplinarity of science and research. Furthermore, the curriculum incorporates a variety of activities and assessment styles that involve reading comprehension, data analysis, and basic math skills. Thus, the lessons also meet many of the Science and Engineering Process Standards laid out by the States of Illinois and Indiana.

The curriculum consists of 10 lesson plans with material for up to twenty-five 45-minute class periods. Early lessons focus on learning the science involved in aquaponics and are lecture/worksheet heavy to deliver the necessary core content. To make the lectures interactive, many brainstorming slides are included to encourage student participation. As the students' knowledge grows, the lessons and activities shift to self-guided learning and project-based activities that require students to apply their knowledge to complex problems.

The lessons are flexible, and educators are encouraged to edit or add content as is seen fit. Lessons were developed to build on each other and be delivered in sequence, but efforts were made so that the lessons can stand alone if educators would like to select only certain lessons.

The graphics work by Joel Davenport, courtesy of Illinois–Indiana Sea Grant, adds a vibrant touch to the curriculum, enhancing its visual and educational appeal.

A NOTE ON USING YOUR SYSTEM

A common roadblock to using your system is the physical space in which the system is set up. In many schools, it is simply not conducive for the large number of students in a class. Finding the best way for students to interact regularly with your system can be tricky, especially if a second staff member is not available to help. If you find yourself in this situation, consider creating groups that rotate through managing and monitoring the system each week. Have those students make observations and report back to the class on the health of the system. Most students get excited about visiting the system, and it is worth finding a way to maximize their time interacting with it.

Lesson 01
Introduction to Aquaponics

- Slide Deck
- Worksheet/Activity
- Teacher's Guide

Lesson 02
Anatomy of an Aquaponics System

- Slide Deck
- Worksheet/Activity
- Teacher's Guide
- Quiz

Lesson 03
Introduction to the Nitrogen Cycle

- Slide Deck
- Worksheet
- Teacher's Guide

Lesson 04
Measuring Nitrogen Levels in Your System

- Slide Deck
- Worksheet/Activity
- Teacher's Guide
- Optional Activity

Lesson 05
Nitrogen Cycle & Population Dynamics

- Slide Deck
- Worksheet
- Teacher's Guide

Lesson 06
Plants in Aquaponic Systems

- Slide Deck
- Worksheet/Activity
- Plant Nutrient Guide
- Teacher's Guide
- Additional Resources

Lesson 07
Fish in Aquaponics Systems

- Slide Deck
- Worksheet/Activity
- Mini Lab
- Teacher's Guide
- Quiz

Lesson 08
Our Modern Food System

- Jigsaw Readings
- Worksheet
- Teacher's Guide

Lesson 09
Troubleshooting Water Quality Challenges

- Worksheet
- Teacher's Guide

Lesson 10
Building an Aquaponics Business

- Slide Deck
- Worksheet
- Teacher's Guide

Appendix
Standards and Alignment

- Primary NGSS
- Additional NGSS
- Great Lakes Literacy

LESSON 1

Introduction to Aquaponics

Teacher's Guide

Time to Complete

45 MINUTES

PREPARATION

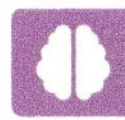

Prerequisite Knowledge

Nutrients

(However, this lesson could incorporate the definition of nutrients and serve as the introduction to these concepts.)

Vocabulary

aquaculture
aquaponics
cages
deep water culture
flood and drain system
hydroponics
net pens
nutrient film technique
raceways
tanks

Learning Objectives

- Students will be able to identify aquaponics, hydroponics, and aquaponics systems.
- Students will be able to describe the basic functions of combining aquaculture and hydroponics systems.
- Students will identify a current issue and reflect on how aquaponics could be used to address the problem.

NGSS Alignment

HS-ETS1-3. Evaluate solution to a complex real-world problem based on prioritized criteria and trade-offs that account for a range of constraints, including cost, safety, reliability, and aesthetics as well as possible social, cultural, and environmental impacts.

Prep Work

- ☐ Print copies of the worksheet and accompanying article for students.
  - ☐ Worksheet: *Introduction to Aquaponics: Hydroponics Used to Feed U.S. Troops*
  - ☐ Answer key at the end of this lesson plan
  - ☐ Article: "National Nutrition Month: Hydroponics Feed Ailing WWII Army Air Forces Personnel," https://www.airforcemedicine.af.mil/News/Article/582803/national-nutrition-month-hydroponics-feed-ailing-wwii-army-air-forces-personnel/
- ☐ If you would like to give this lesson real-world purpose, you can schedule a scientist or farmer to visit your classroom virtually through Illinois–Indiana Sea Grant's Students Ask Scientists (https://iiseagrant.org/education/students-ask-scientists/) and Students Ask Farmers classroom video chat programs. You can schedule a video chat with a scientist online at https://iiseagrant.org/education/students-ask-scientists. To schedule a video chat with a farmer, email iisg@purdue.edu.

Lesson Overview

1. Introduce the concepts of aquaculture and hydroponics to students to provide background information that will help them understand what aquaponics is and how the plants and animals in the system benefit from combining the two culture methods

by going through the definitions in the slide deck, showing videos, and giving the students time to share their own experiences. If you have scheduled a Students Ask Scientists or Students Ask Farmers video chat, allow ample time for discussion.

2. Allow 3 to 5 minutes for students to share their experiences about visiting an aquaculture facility. Ask them follow-up questions. For example:

    a. What did you see?

    b. What kinds of animals were being raised?

    c. Was the farm indoors or outdoors?

    d. What were the animals grown for?

3. The last slide in the slide deck introduces the concept of sustainable aquaculture. Allow 3 to 5 minutes for discussion. Define sustainable aquaculture, then have students answer the question *How does aquaponics fit the definition for sustainable aquaculture?* Answers:

    a. Eliminates pollution (chemical fertilizers, herbicides, pesticides).

    b. Reduces water usage.

    c. Can be used in places where other techniques cannot (for example, dry desert areas or cold climate regions).

    d. Reduces mileage and transportation emissions (local aquaponics farms help reduce the distance food travels from farm to table).

4. End with the assignment *Introduction to Aquaponics: Hydroponics Used to Feed U.S. Troops*. Hand out the printed Introduction to Aquaponics worksheet and the article "National Nutrition Month: Hydroponics Feed Ailing WWII Army Air Forces Personnel." If students prefer to read the article online, you can find it here: https://www.airforcemedicine.af.mil/News/Article/582803/national-nutrition-month-hydroponics-feed-ailing-wwii-army-air-forces-personnel/. This article provides an example of how hydroponics was used in the past to provide nutritious food for American troops. Give students 10 to 15 minutes to read the article and complete the worksheet. Online research can be used if students get stuck. If time permits, have students share the problem they selected in today's world and their thoughts about how aquaponics could be used to solve the problem.

Access the Lesson 1 slide deck by scanning the QR code below or following the link provided.

https://docs.lib.purdue.edu/aquaponics/1

LESSON 1
Introduction to Aquaponics
Worksheet

Introduction to Aquaponics: Hydroponics Used to Feed U.S. Troops

Name ______________________ **Date** __________ **Class Period** _______

Instructions: Read the article "National Nutrition Month: Hydroponics Feed Ailing WWII Army Air Forces" and reflect on how aquaponics could be used to solve a current problem.

1. What was the problem outlined by the article?

2. In a sentence or two, summarize what the solution to the problem was.

3. What is one current problem in the world that aquaponics could be used to solve?

4. Describe how aquaponics could be used to solve the problem.

LESSON 1

Introduction to Aquaponics

Worksheet Answer Key

Introduction to Aquaponics: Hydroponics Used to Feed U.S. Troops

Name ______________________ **Date** __________ **Class Period** ______

Instructions: Read the article "National Nutrition Month: Hydroponics Feed Ailing WWII Army Air Forces" and reflect on how aquaponics could be used to solve a current problem.

1. What was the problem outlined by the article?

   The troops lacked access to healthy, well-balanced meals, and fresh food.

2. In a sentence or two, summarize what the solution to the problem was.

   The U.S. Air Force built hydroponics greenhouses in remote stations to grow vegetables for the troops.

3. What is one current problem in the world that aquaponics could be used to solve?

   - Pollution caused by the use of chemical fertilizers, pesticides, and/or herbicides.
   - Water conservation/limited water resources.
   - Limited growing seasons.
   - Global warming—traditional farms require acres of farmland and produce is often transported over long distances.
   - Food insecurity.

4. Describe how aquaponics could be used to solve the problem.

   - In aquaponics, fish waste is used to feed plants. As a result, the need for chemical fertilizers is eliminated. No chemical waste is released into the environment. There's also no soil, which means no weeds. Growing plants in a soilless system eliminates the need to control weeds. No herbicides are needed.
   - Since the fish feed the plants and the plants clean the water, aquaponics systems can reuse water, which reduces water usage.
   - Food can be grown year-round and close to the communities that need it.
   - Aquaponics farms located near markets can reduce farming's carbon footprint because the produce won't need to be shipped long distances.

LESSON 2

Anatomy of an Aquaponics System

Teacher's Guide

Time to Complete

45–90 MINUTES

(Depending on how long students spend creating a 2D drawing of their system.)

PREPARATION

Prerequisite Knowledge

Abiotic/biotic

(However, this lesson could incorporate the definition of abiotic/biotic factors and serve as the introduction to these concepts.)

Vocabulary

aerator/airstone
biofilter
closed-loop system
dechlorinate
deep water culture
fish tank
grow bed
water cycle
water pump

Learning Objectives

- Students will create a 2D drawing of an aquaponics system and define the role of each of its components.
- Students will identify the components of their onsite aquaponics system and explain the function of each of the components.
- Students will be able to identify the abiotic and biotic factors in their system.

NGSS Alignment

HS-LS2-4. Use mathematical representations to support claims for the cycling of matter and flow of energy among organisms in an ecosystem.

Prep Work

- ☐ Print copies of the worksheet for students.
  - ☐ Worksheet: Getting to Know Your Aquaponics System
  - ☐ Answer key at the end of this lesson plan
- ☐ Gather rulers and yardsticks for student use.
- ☐ Print the *Anatomy of an Aquaponics System* quiz if you choose to use it as an additional assessment tool at the end of this lesson.

Lesson Overview

1. Take students to the aquaponics system. Allow them at least 20 to 30 minutes to explore and create a 2D drawing (model). Bring rulers and yardsticks if you want students to include approximate dimensions in their drawings. As indicated in the student worksheet, complete accuracy is not necessary. The goal of this activity is for students to get to know the different parts of the system, follow the flow of the water, and think critically about the flow of energy through the system and the role of each of the components.
2. Once students have completed their 2D drawings (models), return to the classroom and use the accompanying slide deck to review and discuss student responses to the rest of the questions on the worksheet. Have students label each part of the aquaponics system on their 2D drawing after the terms and functions are discussed.

3. Give students 3 to 5 minutes to fill in the table on page 2 of their worksheet. This can be done as an individual or group activity.
4. As an additional assessment tool, administer the *Anatomy of an Aquaponics System* quiz at the end of Lesson 2 or before beginning Lesson 3.

Additional Resources

- IISG's "Students Ask Scientists: Video Chats": https://iiseagrant.org/education/students-ask-scientists

Access the Lesson 2 slide deck by scanning the QR code below or following the link provided.

https://docs.lib.purdue.edu/aquaponics/2

LESSON 2

Anatomy of an Aquaponics System

Worksheet

Getting to Know Your Aquaponics System

Name ______________________________ **Date** __________ **Class Period** ______

Instructions: Make a 2D drawing of your aquaponics system. Use arrows to show the flow of energy and matter through the system and provide explanations of the flow as needed to support your design. Be sure to indicate the flow of the water as well. For bonus points, include numbers and units of energy to illustrate the flow of energy through the system. The numbers can be hypothetical.

Include: fish tank, grow light, biofilter, grow bed, water pump, airstone, water heater

Be prepared to explain why there are more plants in an aquaponics system in terms of biomass compared to fish.

Name ______________________________ Date __________ Class Period ______

1. Explain how each component below contributes to the function of the entire system.

Fish tank:		Grow bed:	
Grow light:		Water pump:	
Biofilter:		Airstone:	
Water pipes:		Water heater:	

2. Using the terms above in #1, label each component of the system on your drawing.

3. Which components above are abiotic and which are biotic? Provide evidence and reasoning for your claim.

Abiotic	
Biotic	

Name ______________________________ **Date** ___________ **Class Period** _______

4. What inputs are needed to run an aquaponics system? List everything you can think of.

5. What outputs are generated by an aquaponics system?

6. Make a prediction about what will likely happen to the water level in your system over time. What evidence do you have that would support your prediction?

7. How could an aquaponics system be incorporated into your school? What could students learn from the system?

8. How can aquaponics improve current farming methods?

LESSON 2

Anatomy of an Aquaponics System

Quiz

Getting to Know Your Aquaponics System

Name ______________________________ **Date** __________ **Class Period** ______

Instructions: Match the term with the correct definition. Write the number of the definition that matches with the term on the blank line to the right of the term.

Term		Definition
A. Aquaponics	_____	1. space for plants to grow
B. Aquaculture	_____	2. device used to heat water
C. Hydroponics	_____	3. tank where fish live
D. Fish tank	_____	4. farming in water
E. Grow lights	_____	5. mechanical pump that moves water
F. Biofilter	_____	6. process of growing plants without soil
G. Water pipes	_____	7. light used to grow plants
H. Grow bed	_____	8. farming method that combines aquaculture with hydroponics
I. Water pump	_____	9. something that allows air to be blown through it to produce bubbles
J. Airstone	_____	10. any type of filter where bacteria are attached to media
K. Water heater	_____	11. pipes used that transport water

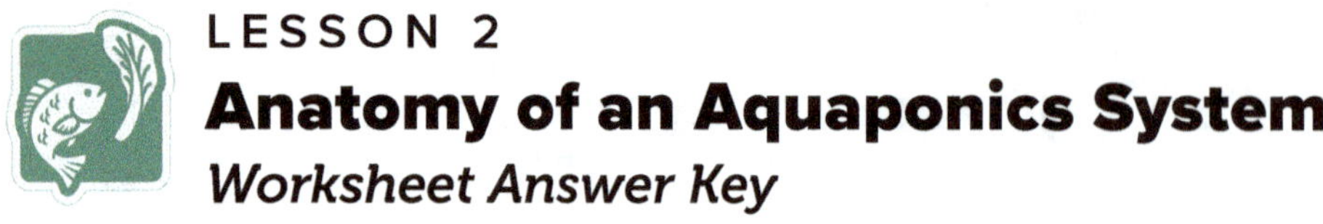

LESSON 2

Anatomy of an Aquaponics System

Worksheet Answer Key

Getting to Know Your Aquaponics System

Name ______________________________ **Date** ___________ **Class Period** ______

Instructions: Make a 2D drawing of your aquaponics system. Use arrows to show the flow of energy and matter through the system and provide explanations of the flow as needed to support your design. Be sure to indicate the flow of the water as well. For bonus points, include numbers and units (kcal/joules) of energy to illustrate the flow of energy through the system. The numbers can be made up.

Include: fish tank, grow light, biofilter, grow bed, water pump, airstone, water heater

Be prepared to explain why there are more plants in an aquaponics system in terms of biomass compared to fish.

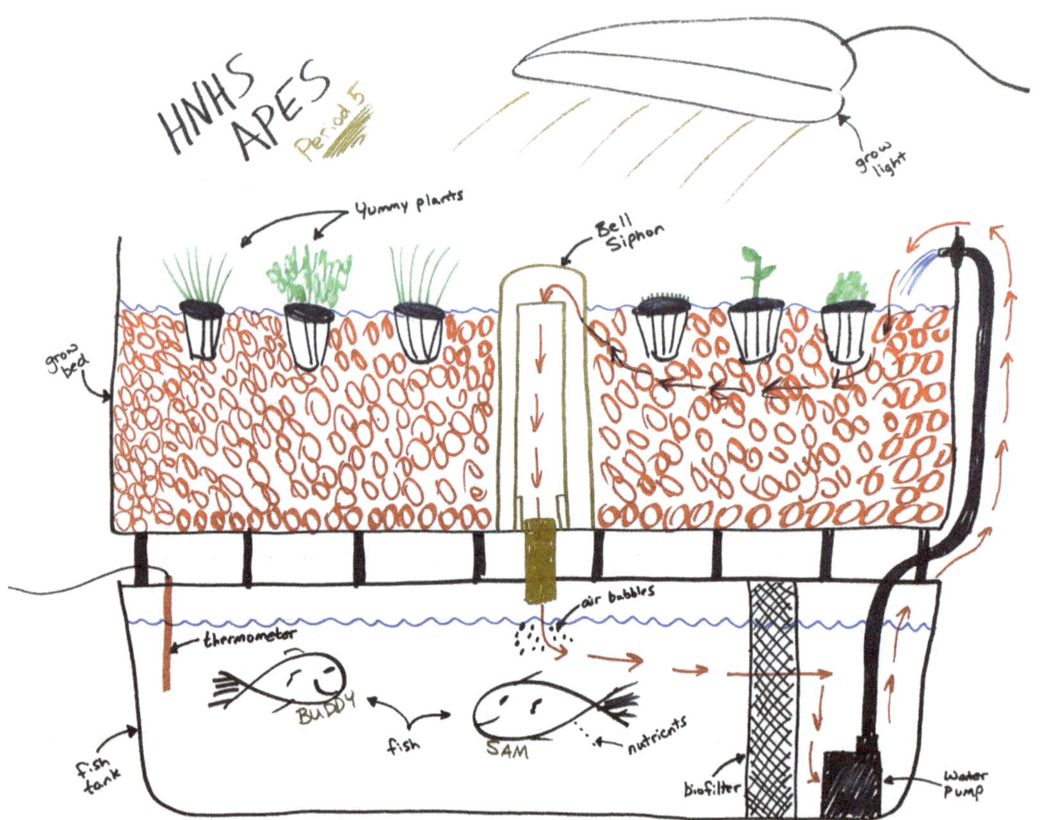

Artwork courtesy of Huntington North High School.

When fish consume fish feed, they use some of the nutrients for energy and excrete what they do not use. Their waste feeds bacteria and plants. The energy that enters the system when fish are fed supports more plants than fish.

Name ______________________________ Date __________ Class Period ______

1. Explain how each component below contributes to the function of the entire system.

Fish tank:	Tank where fish live	Grow bed:	Space for plants to grow; fish tank water flows through the grow bed
Grow light:	Light used to power plant photosynthesis	Water pump:	Mechanical pump that drives water up against gravity to cycle through aquaponics system
Biofilter:	Filters out sediment and has lots of surfaces for bacteria to grow on (not necessary if using media in grow bed)	Airstone:	Connected to pump that adds oxygen to the water, which keeps the fish and plants healthy
Water pipes:	Piping that transports water throughout the system	Water heater:	Heats water to keep fish at healthy temperature (may not be necessary)

2. Using the terms above in #1, label each component of the system on your drawing.

3. Which components above are abiotic and which are biotic? Provide evidence and reasoning for your claim.

Abiotic	Fish tank, grow beds, water pump, grow light, water (nutrients in water, like nitrogen compounds), airstone—oxygen levels, temperature, pH, humidity, air quality
Biotic	Fish, plants, bacteria

Name ______________________________ Date __________ Class Period ______

4. What inputs are needed to run an aquaponics system? List *everything* you can think of.

   - Power for lights, air pump, water pump, water heater
   - Fish food
   - Water (initial tank fill and when water levels drop—very little compared to traditional agriculture)
   - Seedlings
   - Indoor space for the system
   - Maintenance

5. What outputs are generated by an aquaponics system?

   - Fresh, locally sourced vegetables
   - Fresh, locally sourced fish (if you want to harvest them)

6. Make a prediction about what will likely happen to the water level in your system over time. What evidence do you have that would support your prediction?

   As shown in the diagram, the water is in a closed cycle, meaning it flows in a continuous loop through the system. However, water levels slowly drop due to evaporation as well as consumption by plants, and you will eventually need to add water.

7. How could an aquaponics system be incorporated into your school? What could students learn from the system?

   It could provide understanding about plant life cycles and their structure, gardening, issues affecting ecosystems, nutrition, and how to use recycled materials related to sciences and sustainable farming.

8. How can aquaponics improve current farming methods?

   It uses a closed-loop system, conserves water, and offers a sustainable way of producing fish and vegetables.

LESSON 2

Anatomy of an Aquaponics System

Quiz Answer Key

Getting to Know Your Aquaponics System

Name ______________________ **Date** __________ **Class Period** _______

Instructions: Match the term with the correct definition. Write the number of the definition that matches with the term on the blank line to the right of the term.

Term	Answer	Definition
A. Aquaponics	8	1. space for plants to grow
B. Aquaculture	4	2. device used to heat water
C. Hydroponics	6	3. tank where fish live
D. Fish tank	3	4. farming in water
E. Grow lights	7	5. mechanical pump that moves water
F. Biofilter	10	6. process of growing plants without soil
G. Water pipes	11	7. light used to grow plants
H. Grow bed	1	8. farming method that combines aquaculture with hydroponics
I. Water pump	5	9. something that allows air to be blown through it to produce bubbles
J. Airstone	9	10. any type of filter where bacteria are attached to media
K. Water heater	2	11. pipes used that transport water

N
7
Nitrogen

LESSON 3

Introduction to the Nitrogen Cycle

Teacher's Guide

Time to Complete

45 MINUTES

PREPARATION

Prerequisite Knowledge

Introductory knowledge of bacteria

(useful but not necessary)

Introductory knowledge of molecular structure

Vocabulary

aerobic
ammonia
assimilation
bacteria
denitrification
fixation
nitrate
nitrification
nitrite
Nitrobacter
nitrogen
Nitrosomonas
ubiquitous

Learning Objectives

- Students will use a model of the global nitrogen cycle to explain how not all molecules are biologically available and that organisms are designed to process certain molecules.
- Students will use a model of the global nitrogen cycle to determine the flow of nitrogen through organisms and the specific nitrogen molecules involved.
- Students will use the pattern of cycling in the global nitrogen cycle to construct a model of how nitrogen is cycled through the aquaponics system.
- Students will use the model of nitrogen cycling in the aquaponics system to explain how sustainable systems use waste as a resource to recycle material through the system.

NGSS Alignment

HS-LS2-4. Use mathematical representations to support claims for the cycling of matter and flow of energy among organisms in an ecosystem.

Prep Work

- ☐ Print copies of the worksheet for students.
  - ☐ Worksheet: *The Nitrogen Cycle*
  - ☐ Answer key at the end of this lesson plan
- ☐ Decide in advance if you want to include the Connection to Chemistry, Chemistry Crossover slides 8 and 9 of the slide deck.

Lesson Overview

1. Begin slide deck. Start the lesson by playing the first 2 minutes and 24 seconds of the USDA-ARS video *What Is the Nitrogen Cycle?* in slide 4, also found here: https://www.youtube.com/watch?v=xN_c5pr87Ok.
2. When you reach slide 11, which introduces the global nitrogen cycle, hand out worksheets. Allow the students time to complete worksheets either in groups or individually, then resume the presentation, getting answers from students as you proceed through slide 20.

3. The second half of the slide deck is a formal lecture on the nitrogen cycle that takes place in the aquaponics system. There is an optional video link provided in the notes on slide 18 to explain the Haber-Bosch process. Links to additional resources can be found below.

4. The final four worksheet questions are there to get your students to think critically about the importance of nitrogen and how nitrogen cycling is similar and different when comparing aquaponics to traditional agricultural practices. If time allows, use the last four questions to spark discussion after students have time to think about the answers on their own.

Additional Resources

- *What Is the Nitrogen Cycle?*: https://www.youtube.com/watch?v=xN_c5pr87Ok
- Haber-Bosch process (PPT slide 18): https://www.youtube.com/watch?v=c4BmmcuXMu8
- "Nitrogen Transformations in Aquaponic Systems"; includes balanced chemical equations for advanced students (PPT slide 28): https://www2.hawaii.edu/~khanal/aquaponics/nitrogen.html

Access the Lesson 3 slide deck by scanning the QR code below or following the link provided.

https://docs.lib.purdue.edu/aquaponics/3

LESSON 3.

Introduction to the Nitrogen Cycle

Worksheet

The Nitrogen Cycle

Name ______________________________ Date __________ Class Period ______

Analyzing the Nitrogen Cycle Model

Instructions: Study the model of the nitrogen cycle before answering the questions.

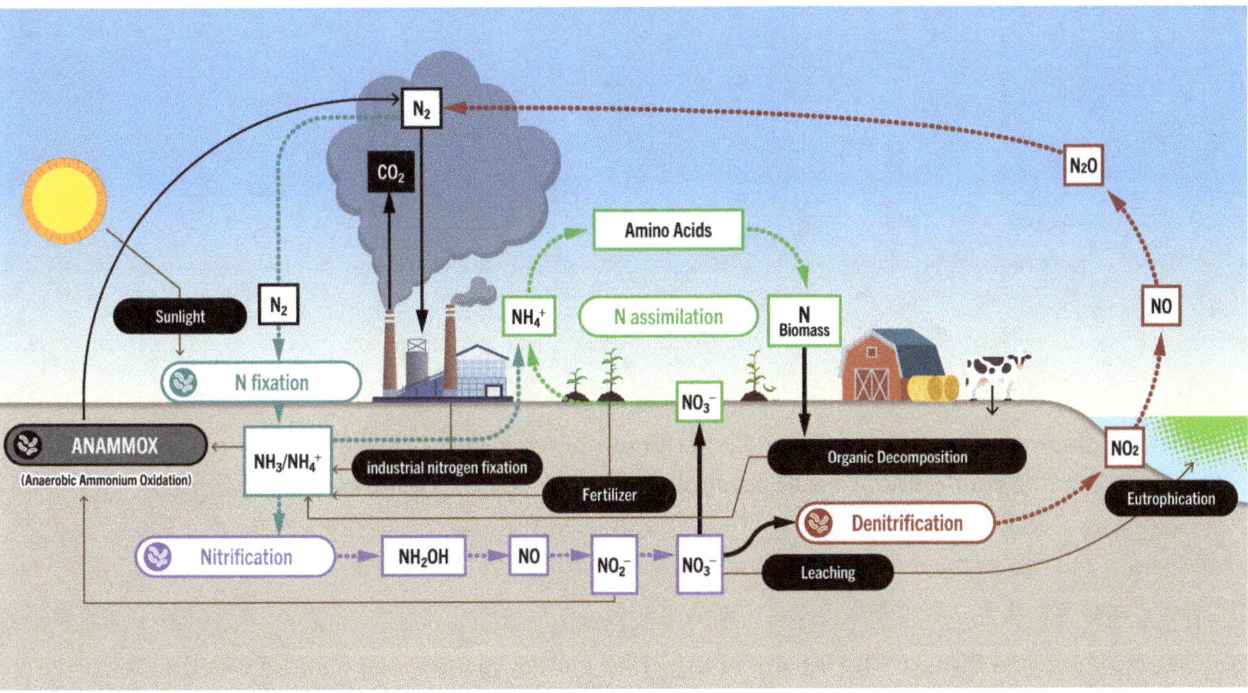

Figure courtesy of Illinois–Indiana Sea Grant.

1. What are all the organisms involved in the nitrogen cycle?

2. On the model, mark how many times bacteria are involved (many different species of bacteria are involved in these processes). How many instances did you find?

Name ______________________________ Date ___________ Class Period _______

3. Starting at the top of the model with N_2, pick one of the outgoing arrows and describe in full sentences how that N_2 is cycled. Include the new chemical species and the organism responsible for the change. You are finished when you return to N_2. (For example, N_2 is converted into fertilizer, NH_3, through an industrial process.)

4. Find and write down all the nitrogen molecules involved in the above model of the nitrogen cycle (N = nitrogen, C = carbon, O = oxygen, H = hydrogen, S = sulfur, P = phosphorous).

5. Make a claim about whether a chemical or physical change is occurring in the nitrogen cycle model. What is your evidence and reasoning to support your claim?

6. Seeing the cycle through the lens of energy and matter, make a claim about whether nitrogen is ever created or destroyed in the cycle. What is your evidence and reasoning to support your claim?

7. Humans aren't explicitly in this figure, but their presence is implied. Explain how humans are involved in the global nitrogen cycle.

Name ______________________ Date __________ Class Period ______

8. Why is nitrogen so important for living things?

9. Explain how the nitrogen cycle in the aquaponics system is similar to the global nitrogen cycle.

10. Based on the aquaponics nitrogen cycle, how do nitrogen and other nutrients enter the aquaponics ecosystem over a given amount of time? Include the phrase *nutrient loading* in your answer. Nutrient loading is the amount of nutrients that enter an ecosystem over a given period of time.

11. Thinking about traditional agriculture, which often uses chemical fertilizers, how is this different from aquaponics?

LESSON 3
Introduction to the Nitrogen Cycle
Worksheet Answer Key

The Nitrogen Cycle

Name ______________________ Date __________ Class Period ______

Analyzing the Nitrogen Cycle Model

Instructions: Study the model of the nitrogen cycle before answering the questions.

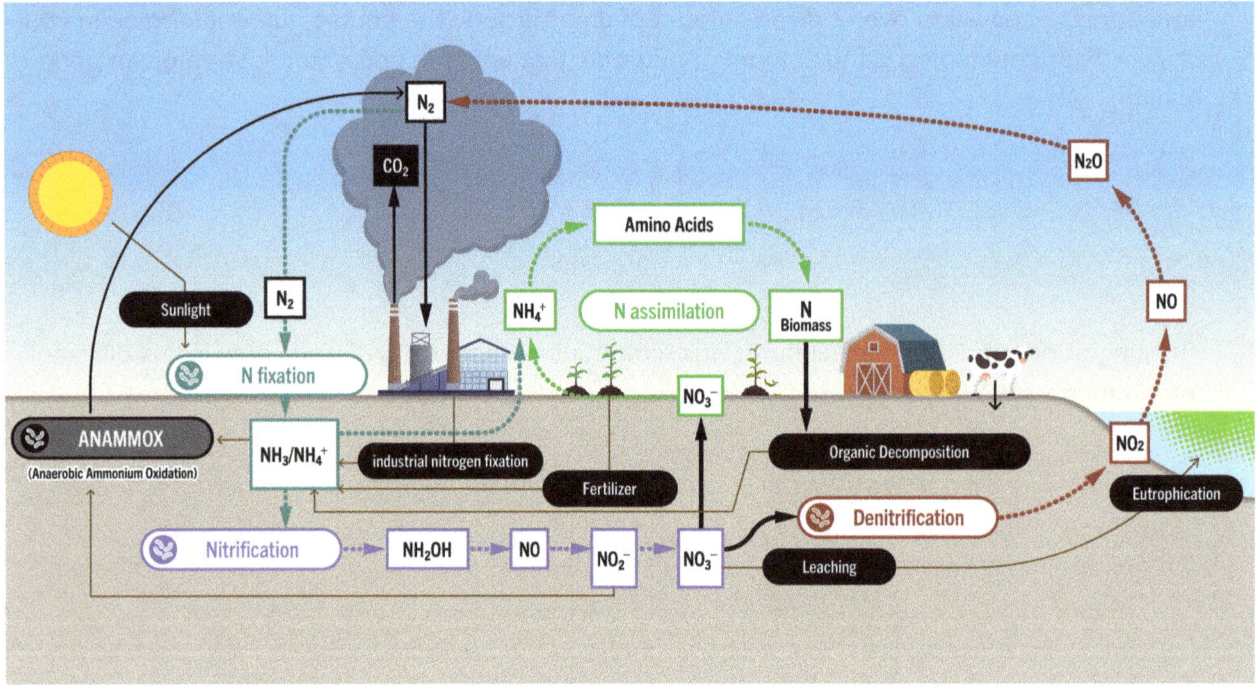

Figure courtesy of Illinois-Indiana Sea Grant.

Study the model of the nitrogen cycle.

1. What are all the organisms involved in the nitrogen cycle?

   Bacteria, plants, animals

2. On the model, mark how many times bacteria are involved (many different species of bacteria are involved in these processes). How many instances did you find?

   Bacteria are involved 4 times

Name ______________________________ Date __________ Class Period ______

3. Starting at the top of the model with N_2, pick one of the outgoing arrows and describe in full sentences how that N_2 is cycled. Include the new chemical species and the organism responsible for the change. You are finished when you return to N_2. (For example, N_2 is converted into fertilizer, NH_3, through an industrial process.)

There are a number of answers possible. Student's answers should start by describing how atmospheric nitrogen (N_2) is converted to ammonia by the process of either industrial or biological nitrogen fixation and end back at N_2.

4. Find and write down all the nitrogen molecules involved in the above model of the nitrogen cycle (N = nitrogen, C = carbon, O = oxygen, H = hydrogen, S = sulfur, P = phosphorous).

- Nitrogen—N_2
- Ammonia/ammonium—NH_3/NH_{4^+}
- Nitrite—NO_{2^-}
- Nitrate—NO_{3^-}
- Nitric oxide—NO
- Hydroxylamine—NH_2OH
- Nitrous oxide—N_2O

5. Make a claim about whether a chemical or physical change is occurring in the nitrogen cycle model. What is your evidence and reasoning to support your claim?

Chemical change is taking place.

6. Seeing the cycle through the lens of energy and matter, make a claim about whether nitrogen is ever created or destroyed in the cycle. What is your evidence and reasoning to support your claim?

No, it is never created nor destroyed. Conservation of mass is always at play!

7. Humans aren't explicitly in this figure, but their presence is implied. Explain how humans are involved in the global nitrogen cycle.

Agriculture, fertilizer, livestock—could get more abstract here and hypothesize that because humans are impacting climate, we will inevitably have an effect on the entire system.

Name ______________________________ Date __________ Class Period ______

8. Why is nitrogen so important for living things?

Nitrogen is a key element in the nucleic acids DNA and RNA, which are the most important of all biological molecules and crucial for all living things.

9. Explain how the nitrogen cycle in the aquaponics system is similar to the global nitrogen cycle.

The aquaponics nitrogen cycle is a subset of the global nitrogen cycle.

10. Based on the aquaponics nitrogen cycle, how do nitrogen and other nutrients enter the aquaponics ecosystem over a given amount of time? Include the phrase *nutrient loading* in your answer. Nutrient loading is the amount of nutrients that enter an ecosystem over a given period of time.

Nutrient loading occurs in aquaponics when fish feed is added to the system. Fish are fed daily, allowing for nitrogen and other nutrients to be continually loaded into the system.

11. Thinking about traditional agriculture, which often uses chemical fertilizers, how is this different from aquaponics?

In aquaponics, the fertilizer for the plants is the nitrogenous waste from fish. Chemical fertilizers are not needed or used.

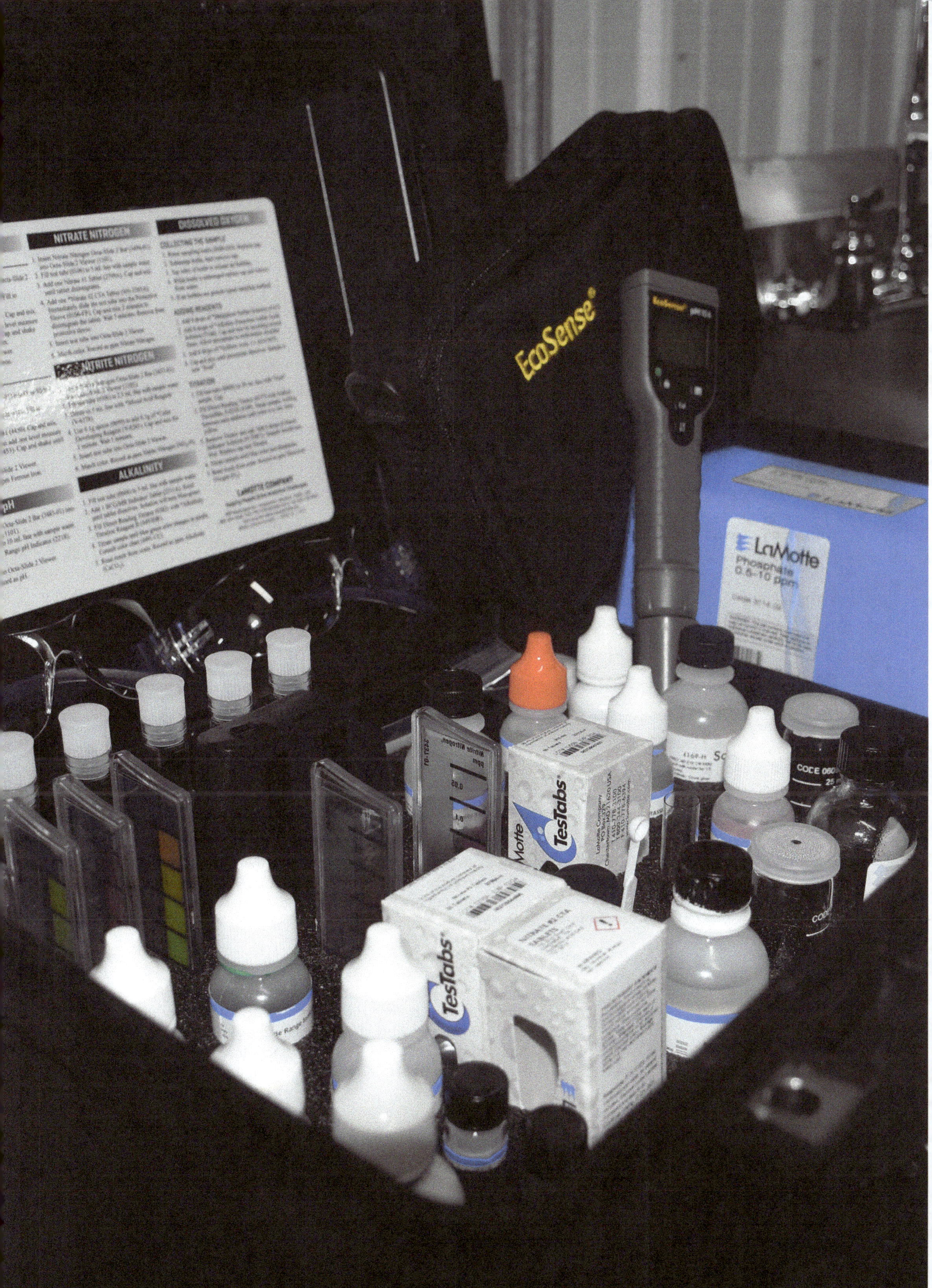

NITRATE NITROGEN
DISSOLVED OXYGEN
NITRITE NITROGEN
ALKALINITY
pH
EcoSense
LaMotte
Phosphate
0.5-10 ppm
TesTabs
TesTabs
Nitrite Nitrogen

LESSON 4

Measuring Nitrogen Levels in Your System

Time to Complete

1–3 CLASS PERIODS

(Depending on how much time is spent on water testing.)

PREPARATION

Prerequisite Knowledge

Familiar with conversion calculations (useful but not required)

Introductory knowledge of molecular structure and Avogadro's number

Vocabulary

concentration
dissolved oxygen (DO)
homogeneous
intensive property
molarity
molecular weight
parts per million (ppm)
pH
toxicity

Learning Objectives

- Students will test the water in the aquaponics system for pH, temperature, and concentrations of ammonia, nitrite, nitrate, and dissolved oxygen (DO).
- Students will perform calculations to convert between ppm and molarity.
- Students will compare their measurements and calculations to the levels that create toxicity—and the concentrations at which certain molecules/elements are toxic—to ensure that their aquaponics system concentrations remain at healthy levels.

NGSS Alignment

HS-LS2-4. Use mathematical representations to support claims for the cycling of matter and flow of energy among organisms in an ecosystem.

Prep Work

- ☐ Print copies of the worksheet for students.
  - ☐ Worksheet: *Calculating and Converting Concentrations*
  - ☐ Answer key at the end of this lesson plan
- ☐ Print copies of *Know Your H2O Quick Reference Guide for Testing Water Quality* if you have decided to use it as a supplemental resource.
- ☐ Decide in advance if you want to use the optional writing assignment, *The Journey of Nitrogen Through Your Aquaponics System*, to get your students' creative thinking juices flowing and to reinforce concepts learned in Lessons 3 and 4. If you decide to use it, print out the assignment in advance.

Lesson Overview

1. Hand out the worksheet *Calculating and Converting Concentrations* before you start the lesson. Talk through the slide deck, pausing to allow students time to do calculations. (If you don't want to focus so heavily on concentration measurements, omit the worksheet and cover the calculations solely in the presentation.)

2. Once finished with the slide deck, move on to learning how to perform the water tests. This can often be a logistical challenge with so many students and, usually, too few water testing kits. We recommend two options for water testing:

    a. demonstrating the testing process for the whole class yourself with the students writing down instructions to keep them engaged, or

    b. creating "expert" groups for each test. In this second option, group students by test (ammonia, nitrite, nitrate, and pH) and have them learn that specific test.

    Testing equipment and supplies can be borrowed from Illinois-Indiana Sea Grant through the Know Your H2O loan program (https://iiseagrant.org/education/loanable-kits/). Dissolved oxygen meters, pH pens, and water chemistry testing kits can be purchased online from distributors or directly from the manufacturers. Test kits can also be purchased from pet supply stores that have an aquarium section.

3. It is often interesting for students to test their tap water in addition to their system water to confirm the safety of their own drinking water.

4. Once testing is complete, have students share the results with the class. Moving forward, for future testing, these students will be the experts on those tests.

5. IMPORTANT NOTES:

    a. Read water testing instructions carefully and stress to your students that following them exactly is critical—in particular, shaking the solutions and timing.

    b. For water testing after Lesson 4: The proper approach to monitoring nitrogen levels in your system is to test weekly, if possible, once the nitrogen cycle is complete. Testing more frequently when starting your system, and graphing the results, is a good way to reinforce concepts learned in Lesson 3.

    Once students have learned how to test, create a routine where different groups of students test the water weekly. For example, a group of students tests the water levels every Monday. It might be best for you to collect the water and have it in your classroom if you don't want students walking down to the system by themselves. If you created experts in the water testing lesson, combine them to create weekly groups that include one or more experts for each test. Once they have finished testing, the students can add their values to the record sheet and present their results to the class, discussing any significant changes and the overall health of the system. This is also a good time to check the water level and top up if necessary.

6. If you have decided to use the optional writing assignment to get your students' creative thinking juices flowing and to reinforce concepts learned in Lesson 3 and 4, hand out the assignment *The Journey of Nitrogen Through Your Aquaponics System* to complete in class or as homework.

Additional Resources

- Know Your H2O water quality test kit loan program: https://iiseagrant.org/education/loanable-kits/
- *Know Your H2O Quick Reference Guide for Testing Water Quality*: https://iiseagrant.org/wp-content/uploads/2024/10/Know-Your-H2O-Quickstart-Guide-11.pdf
- *Important Water Quality Parameters in Aquaponics Systems*: https://pubs.nmsu.edu/_circulars/CR680/

Access the Lesson 4 slide deck by scanning the QR code below or following the link provided.

https://docs.lib.purdue.edu/aquaponics/4

LESSON 4

Measuring Nitrogen Levels in Your System

Worksheet

Calculating and Converting Concentrations

Name ______________________ **Date** __________ **Class Period** ______

Instructions: Answer the questions below. Write down all steps in your calculation. Circle your answers. Don't forget to include units.

$$\text{ppm} = \frac{\text{mass of solute}}{\text{mass of solution}} \times 1{,}000{,}000$$

1. A 12 oz can of Coke contains 39 grams of sugar. What is the concentration of sugar in a 12 oz can of Coke in ppm? 12 oz of Coke is 369 grams.

2. How many nitrite molecules are in a 10-gallon tank if the concentration is 1 ppm? To answer this question, you will need to use all of the equivalencies provided.

   1 mL aquarium water = 0.00026 gallons aquarium water
   1 mol nitrite = 6.022×10^{23} molecules of nitrite

   If you're confused, work to cancel out the units until you're left with molecules of nitrite (it will be a big number). Here's the first step:

$$\text{ppm} = \frac{1\,\text{g nitrite}}{1{,}000{,}000\ \cancel{\text{g aquarium water}}} \times \frac{1\,\cancel{\text{g aquarium water}}}{1\ \text{mL aquarium water}}$$
$$= \frac{1\,\text{g nitrite}}{1{,}000{,}000\ \text{mL aquarium water}}$$

Name ______________________________ Date __________ Class Period ______

3. Once you have measured the nitrate concentration, estimate the total number of nitrate molecules in your system. If not already known, you will need to approximate the total water volume in your system.

4. Based on your measurements and calculations of nitrite and nitrate, what do you know about the nitrogen cycle in your systems?

5. Are the levels of ammonia safe for your fish?

6. Are the levels of nitrite safe for your fish?

7. Are the levels of nitrate safe for your fish?

LESSON 4

Measuring Nitrogen Levels in Your System

Optional Activity

The Journey of Nitrogen Through Your Aquaponics System

Name ______________________________ **Date** __________ **Class Period** ______

Instructions: Answer the question below:

Imagine that you are a nitrogen molecule in the aquaponics system at your school. Write a short story about the journey you take as you travel through the system and the nitrogen cycle. Have fun with it!

Include the following terms in your story:

- ☐ **Nitrogen**
- ☐ **Bacteria**
- ☐ **Fixation**
- ☐ **Fertilizer**
- ☐ **Nitrite**
- ☐ **Nitrate**

LESSON 4

Measuring Nitrogen Levels in Your System

Worksheet Answer Key

Calculating and Converting Concentrations

Name ______________________ **Date** __________ **Class Period** ______

Instructions: Answer the questions below. Write down all steps in your calculation. Circle your answers. Don't forget to include units.

$$\text{ppm} = \frac{\text{mass of solute}}{\text{mass of solution}} \times 1{,}000{,}000$$

1. A 12 oz can of Coke contains 39 grams of sugar. What is the concentration of sugar in a 12 oz can of Coke in ppm? 12 oz of Coke is 369 grams.

$$\frac{39\text{ g sugar}}{369\text{ g solution}} \times 1{,}000{,}000 = 105{,}691\text{ ppm}$$

2. How many nitrite molecules are in a 10-gallon tank if the concentration is 1 ppm? To answer this question, you will need to use all of the equivalencies provided.

1 mL aquarium water = 0.00026 gallons aquarium water

1 mol nitrite = 6.022×10^{23} molecules of nitrite

If you're confused, work to cancel out the units until you're left with molecules of nitrite (it will be a big number). Here's the first step:

$$\text{ppm} = \frac{1\text{ g nitrite}}{1{,}000{,}000\text{ g aquarium water}} \times \frac{1\text{ g aquarium water}}{1\text{ mL aquarium water}}$$

$$= \frac{1\text{ g nitrite}}{1{,}000{,}000\text{ mL aquarium water}}$$

1 mL aquarium water = 0.00026 gallons aquarium water

1 mol nitrite = 6.022×10^{23} molecules of nitrite

$$\frac{1\text{ g nitrite}}{1{,}000{,}000\text{ mL aquarium water}} \times \frac{1\text{ mL aquarium water}}{0.00026\text{ gallons of aquarium water}} \times 10\text{ gallons} = 0.038\text{ g nitrite}$$

$$0.038\text{ g nitrite} \times \frac{1\text{ mol}}{46\text{ g nitrite}} \times \frac{6.022 \times 10^{23}\text{ molecules of nitrite}}{1\text{ mol}} = 4.97 \times 10^{20}\text{ molecules of nitrite}$$

Name ______________________ Date __________ Class Period ______

3. Once you have measured the nitrate concentration, estimate the total number of nitrate molecules in your system. If not already known, you will need to approximate the total water volume in your system.

   Answers will vary based on nitrate test results. Calculations will be similar to those in question 2 except:

   62 grams of nitrate = 1 mol nitrate.

   1 mol of nitrate = 6.022 × 1023 molecules of nitrate.

4. Based on your measurements and calculations of nitrite and nitrate, what do you know about the nitrogen cycle in your systems?

   Answers will vary based on measurements and calculations. Answers should indicate that students understand that nitrite and nitrate in the water is evidence for nitrification.

5. Are the levels of ammonia safe for your fish?

   0 ppm–0.9 ppm of ammonia—Yes, safe.

   ≥1 ppm of ammonia—No, not safe.

6. Are the levels of nitrite safe for your fish?

   The answer will depend on nitrite level and species.

   0 ppm–0.9 ppm of nitrite—Yes, safe.

   ≥1 ppm of nitrite—No, not safe.

7. Are the levels of nitrate safe for your fish?

   The answer will depend on nitrate level and species.

   20–50 ppm of nitrate—Yes, safe.

   ≥50 ppm of nitrate—No, not safe.

LESSON 4

Measuring Nitrogen Levels in Your System

Optional Activity Answer Key

The Journey of Nitrogen Through Your Aquaponics System

Name ______________________________ **Date** __________ **Class Period** ______

Instructions: Answer the question below:

Imagine that you are a nitrogen molecule in the aquaponics system at your school. Write a short story about the journey you take as you travel through the system and the nitrogen cycle. Have fun with it!

Include the following terms in your story:

- ☐ **Nitrogen**
- ☐ **Bacteria**
- ☐ **Fixation**
- ☐ **Fertilizer**
- ☐ **Nitrite**
- ☐ **Nitrate**

Each student's story will be different.

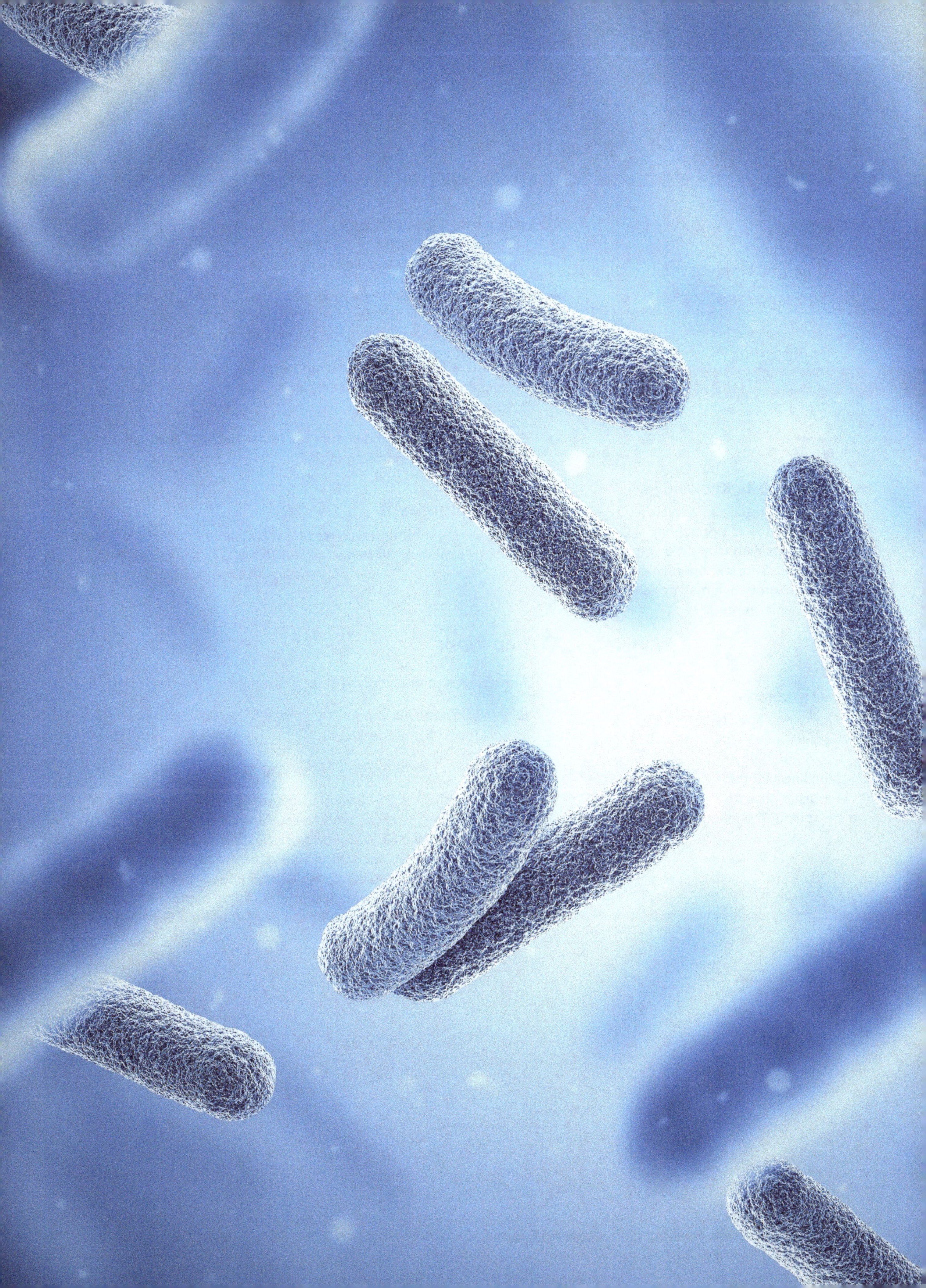

LESSON 5

Nitrogen Cycle & Population Dynamics

Teacher's Guide

Time to Complete

90 MINUTES

PREPARATION

Prerequisite Knowledge

The slide deck has a basic introduction to population dynamics with straightforward definitions for exponential and logistic growth. It may need to be supplemented.

Vocabulary

carrying capacity
equilibrium
exponential growth
limiting factors
logistic growth
population dynamics

Learning Objectives

- Students will apply population dynamics to different organisms.
- Students will calculate exponential growth, limiting factors, carrying capacity, logistic growth, and equilibrium by modeling the population of bacteria in their aquaponics system.
- Students will calculate how changes in temperature and pH would impact the growth curve of a bacteria colony in their aquaponics system.
- Students will explain how changes to the bacteria population would impact the nitrogen cycle.

NGSS Alignment

HS-LS2-2. Use mathematical representations to support and revise explanations based on evidence about factors affecting biodiversity and populations in ecosystems of different scales.

Prep Work

- ☐ Print copies of the worksheet for students.
  - ☐ Worksheet: *Population Dynamics of* Nitrosomonas *in a NEW Aquaponics System*
  - ☐ Answer key at the end of this lesson plan
- ☐ Decide in advance if you want to create your own cycling data for the last two pages of the worksheet *Population Dynamics of* Nitrosomonas *in a NEW Aquaponics System.* If you choose to do this optional activity, you will need a small fishbowl and ammonium chloride solution (fishless cycling) *or* one goldfish and fish food, along with gravel or bioballs and an airstone. You may also want to use water and bioballs from your established system.

Lesson Overview

1. The slide deck begins with figures illustrating the population dynamics in the United States and Indiana. The goal is for students to (a) interpret the trends in the graphs and (b) gain an appreciation for why population dynamics is an important area of research. Encourage them to brainstorm why it's important to predict how the population changes by age on slide 6. These metrics directly impact the size of the labor force, how many schools will be required, predictions for the birth rate, and ultimately how resources (money) will be allocated. They should develop an understanding that modeling the state population by different metrics (e.g., age, location) is relevant to their lives.

2. The slide deck then moves into a basic approach to understanding population dynamics. This culminates in assessing the carrying capacity of fish within the system, which is very simple but will be a good exercise to prepare them for the worksheet.

3. Pass out the worksheet and leave the final slide showing the aquaponics nitrogen cycle up for reference.

4. **Optional Activity:** Time permitting, you could create your own cycling data for the last two pages of the worksheet. This could be done with a small fishbowl and ammonium chloride solution for a fishless cycling or with a goldfish and some fish food, along with gravel or bioballs for bacteria growth and an airstone. The process can be expedited by using water or bioballs from your established system to quickly introduce *Nitrosomonas* and *Nitrobacter* into the system. This would require measuring nitrogen levels every 2 to 3 days for a few weeks, giving students plenty of practice taking measurements. This data gathering and real-time analysis creates a more genuine experience for students but is also time-consuming and potentially costly.

Access the Lesson 5 slide deck by scanning the QR code below or following the link provided.

https://docs.lib.purdue.edu/aquaponics/5

LESSON 5

Nitrogen Cycle and Population Dynamics

Worksheet

Population Dynamics of *Nitrosomonas in a NEW Aquaponics System*

Name ______________________ **Date** __________ **Class Period** ______

Instructions: Imagine that 100 *Nitrosomonas* bacteria find their way to your new aquaponics system, settle, and begin to reproduce all at the same time via binary fission every 7 hours. Plot their population growth for their first 42 hours in the tank. Assume that they have all the resources they need to continue producing at the same rate. Make sure to label your axes.

1. Determine and plot *Nitrosomonas* population growth:

Hour	*Nitrosomonas Population*
0	100
7	
14	
21	
28	
35	
42	

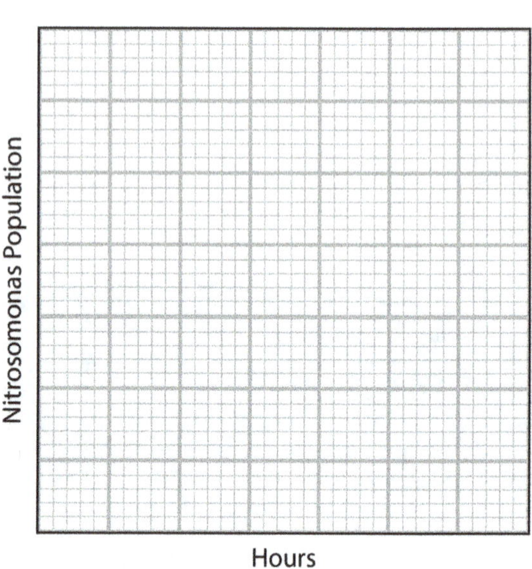

2. You have just drawn an exponential growth curve. Would you expect *Nitrosomonas* to continue reproducing at the same rate forever in the aquaponics system? What are the limiting factors for the population size of *Nitrosomonas*?

Name ______________________________ Date __________ Class Period ______

3. What assumptions are we making about the population dynamics of the *Nitrosomonas* population? Which contributions are we ignoring? (Consider I, E, B, and D.)

4. Now, based on the limiting factors you've named for the *Nitrosomonas* population, assume that the carrying capacity for the system is 5,000 bacteria. (This number is arbitrary and just used to illustrate the concepts). Draw and label the carrying capacity on your graph and add a line to show how the growth curve would change—this is an S-shaped curve.

5. If more fish were added to the system, how would you expect that to affect the carrying capacity for *Nitrosomonas* in the system?

Name ______________________________ Date ___________ Class Period _______

6. Imagine the heat in the school shut down and the temperature dropped right when you started a new aquaponics system. The optimal temperature for bacteria growth is between 77°F and 86°F and the temperature is now 70 F. Which graph do you think best represents how this temperature change would impact the population growth of *Nitrosomonas*? Explain your reasoning.

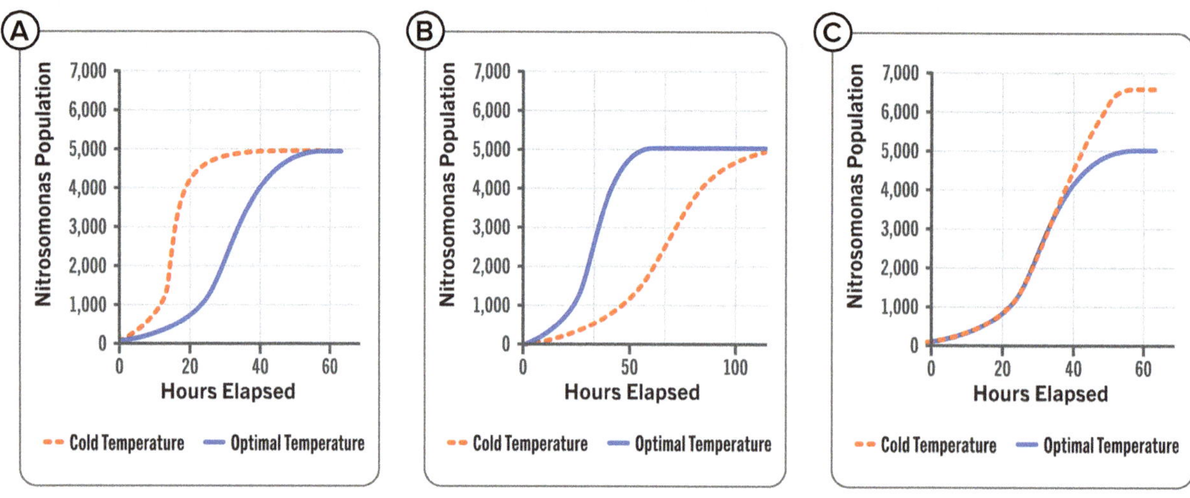

Graphs courtesy of Illinois–Indiana Sea Grant.

7. Now, we will consider all the nitrogen levels as a new aquaponics system is started. The process of starting a fish tank is called *cycling* and begins by adding an ammonia source to your system, either from live fish or by manually adding ammonia (it can be dangerous to cycle with fish because of the high levels of ammonia and nitrite reached during the process.) Again, while we are not directly measuring the population of bacteria in the system, by measuring the concentration of ammonia, nitrite, and nitrate, we can draw conclusions about their arrival and population growth.

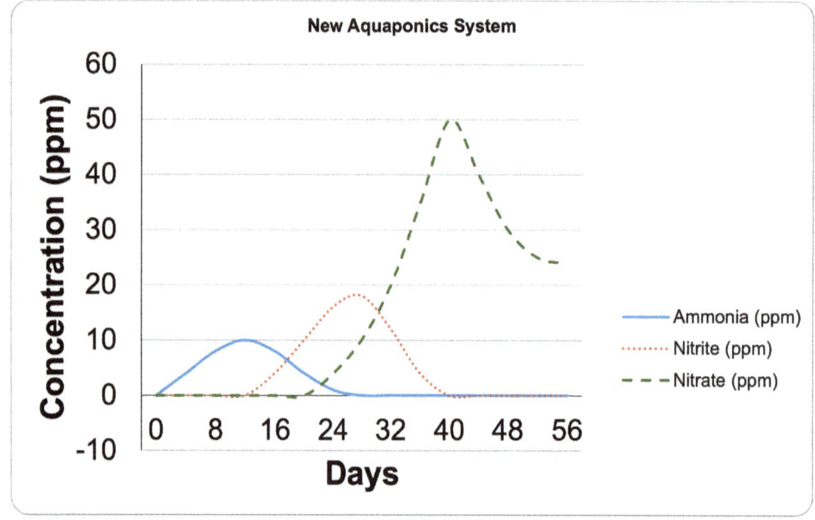

Graph courtesy of Illinois–Indiana Sea Grant.

Name ______________________________ Date __________ Class Period ______

8. Using the graph, write a CER (claim, evidence, reasoning) for when *Nitrosomonas* and *Nitrobacter* arrived and colonized the system.

***Nitrosomonas* Arrival**

Claim:

Evidence:

Reasoning:

***Nitrobacter* Arrival**

Claim:

Evidence:

Reasoning:

9. What happened around day 40? What made the nitrate levels drop in the aquaponics system?

10. Your aquaponics system should (hopefully) have a low or zero concentration of ammonia and nitrite. Does this mean that there is never any ammonia or nitrite in the system? Explain using what's happening using the word *equilibrium*, which means when two opposing forces are balanced.

Name ______________________________ Date ___________ Class Period _______

11. If you measured non-zero concentrations of ammonia or nitrite, what might be going wrong with your system?

12. When will we be able to tell if our system start-up worked?

LESSON 5

Nitrogen Cycle and Population Dynamics

Worksheet Answer Key

Population Dynamics of *Nitrosomonas in a NEW Aquaponics System*

Name ______________________ **Date** __________ **Class Period** ______

Instructions: Imagine that 100 *Nitrosomonas* bacteria find their way to your new aquaponics system, settle, and begin to reproduce all at the same time via binary fission every 7 hours. Plot their population growth for their first 42 hours in the tank. Assume that they have all the resources they need to continue producing at the same rate. Make sure to label your axes.

1. Determine and plot *Nitrosomonas* population growth:

Hour	*Nitrosomonas Population*
0	100
7	200
14	400
21	800
28	1,600
35	3,200
42	6,400

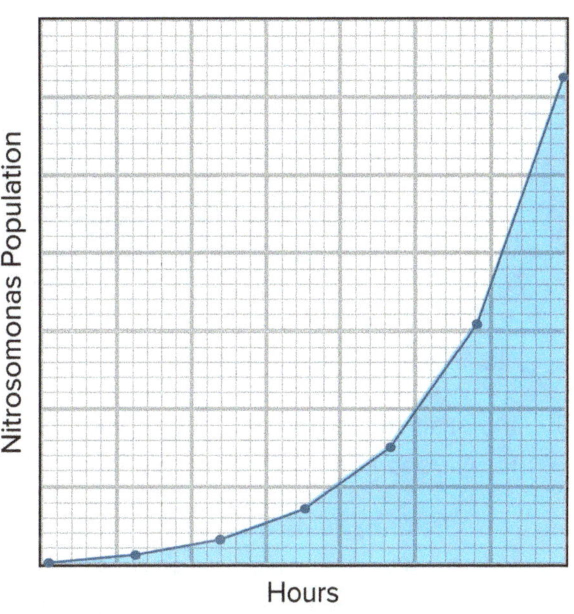

2. You have just drawn an exponential growth curve. Would you expect *Nitrosomonas* to continue reproducing at the same rate forever in the aquaponics system? What are the limiting factors for the population size of *Nitrosomonas*?

No. The bacteria will eventually be limited by the amount of ammonia in the system produced by the fish. Appropriate space for growth could potentially be a limiting factor as well.

3. What assumptions are we making about the population dynamics of the *Nitrosomonas* population? Which contributions are we ignoring? (Consider I, E, B, and D.)

In our current model, we are ignoring new immigration, emigration, and any deaths. We are assuming that the initial immigration and subsequent binary fission (births) are the only contributors to changes in the *Nitrosomonas* population.

4. Now, based on the limiting factors you've named for the *Nitrosomonas* population, assume that the carrying capacity for the system is 5,000 bacteria. (This number is arbitrary and just used to illustrate the concepts.) Draw and label the carrying capacity on your graph and add a line to show how the growth curve would change—this is an S-shaped curve.

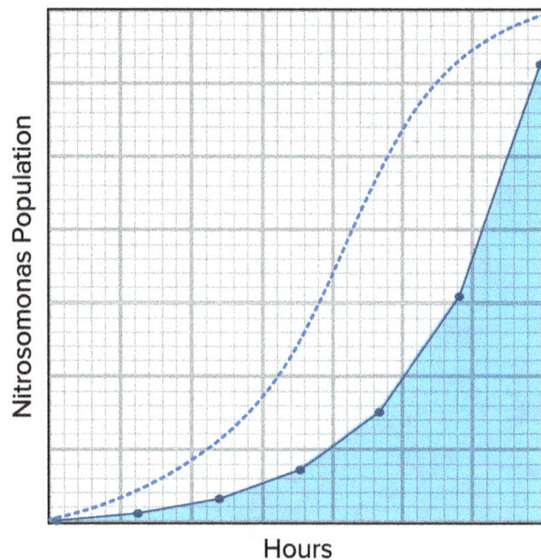

Note: Carrying capacity is 5,000 bacteria.

5. If more fish were added to the system, how would you expect that to affect the carrying capacity for *Nitrosomonas* in the system?

We would expect the carrying capacity to increase if more fish were added because more ammonia would be produced, which is a necessary resource for the *Nitrosomonas* bacteria and was previously a limiting factor.

Alternatively, if there was not enough space for the bacteria population to continue growing, we would expect the carrying capacity to stay the same and the concentration of ammonia to rise in the system.

6. Imagine the heat in the school shut down and the temperature dropped right when you started a new aquaponics system. The optimal temperature for bacteria growth is between 77°F and 86°F and the temperature is now 70°F. Which graph do you think best represents how this temperature change would impact the population growth of *Nitrosomonas*? Explain your reasoning.

The second graph best represents how a temperature change would impact the population growth of Nitrosomonas.

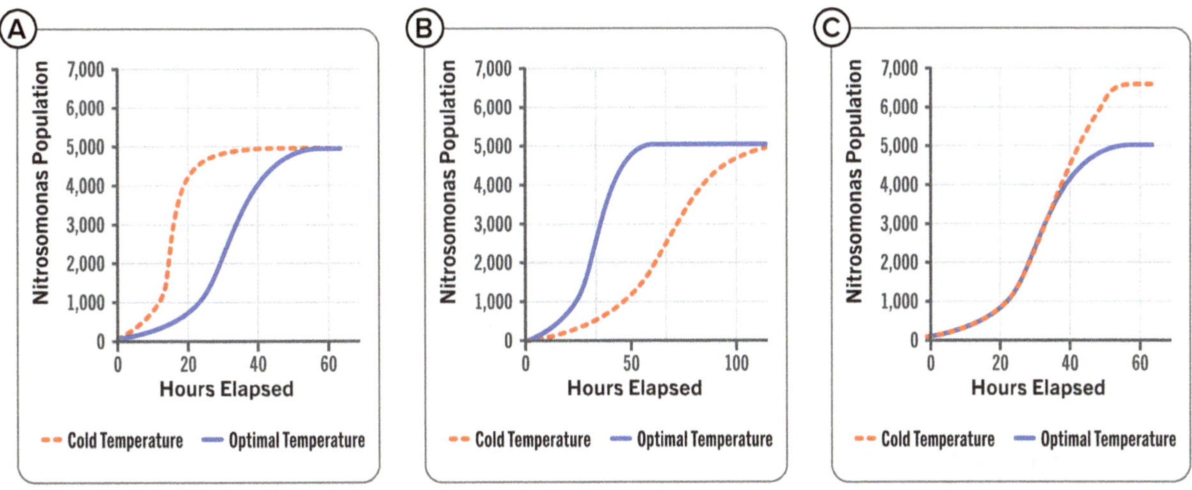

Graphs courtesy of Illinois-Indiana Sea Grant

7. Now, we will consider all the nitrogen levels as a new aquaponics system is started. The process of starting a fish tank is called *cycling* and begins by adding an ammonia source to your system, either from live fish or by manually adding ammonia (it can be dangerous to cycle with fish because of the high levels of ammonia and nitrite reached during the process.) Again, while we are not directly measuring the population of bacteria in the system, by measuring the concentration of ammonia, nitrite, and nitrate, we can draw conclusions about their arrival and population growth.

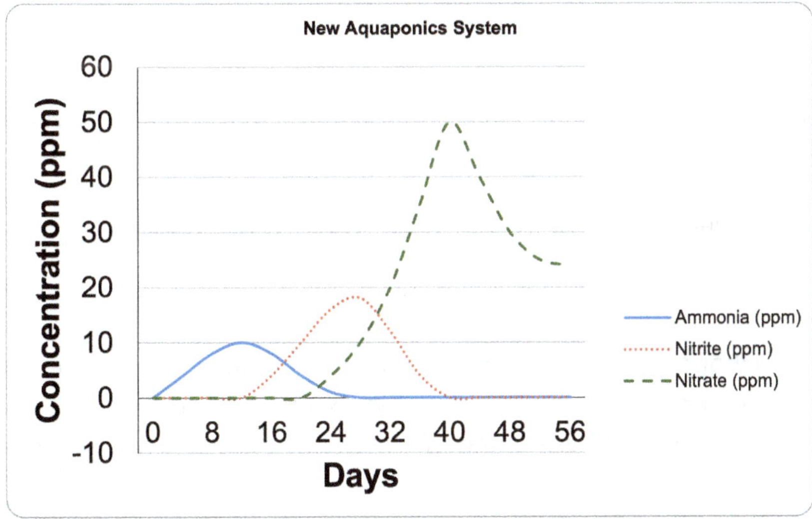

Graph courtesy of Illinois-Indiana Sea Grant

8. Using the graph, write a CER (claim, evidence, reasoning) for when *Nitrosomonas* and *Nitrobacter* arrived and colonized the system.

Nitrosomonas Arrival

Claim:

Nitrosomonas arrived in the system around day 11.

Evidence:

The ammonia concentration begins to drop around day 11.

The nitrite concentration begins to rise around day 11.

Reasoning:

For the ammonia concentration to decrease, *Nitrosomonas* must have established in the system because no other organism in the system consumes ammonia. Similarly, for the nitrite concentration to increase, *Nitrosomonas* must be in the system because it is the only organism to produce it.

Nitrobacter Arrival

Claim:

Nitrobacter arrived in the system around day 20.

Evidence:

The nitrate concentration in the system begins to increase on day 20.

Reasoning:

Nitrobacter is the only organism that produces nitrate, so it must be in the system around day 20, even though the nitrite concentration doesn't start to decrease until day 25.

9. What happened around day 40? What made the nitrate levels drop in the aquaponics system?

Plants were likely introduced to the system on day 40.

10. Your aquaponics system should (hopefully) have a low or zero concentration of ammonia and nitrite. Does this mean that there is never any ammonia or nitrite in the system? Explain using what's happening using the word *equilibrium*, which means when two opposing forces are balanced.

No. Ammonia and nitrite are being created in the system but are immediately consumed by the *Nitrosomonas* and *Nitrobacter*. In other words, the populations of fish, *Nitrosomonas*, and *Nitrobacter* are in equilibrium because the bacteria populations can immediately consume the ammonia and nitrite generated.

11. If you measured non-zero concentrations of ammonia or nitrite, what might be going wrong with your system?

Ultimately, the problem is that the bacteria populations are inadequate for the amount of ammonia or nitrite being produced. This could be caused by an increase in the amount of ammonia/nitrite produced (more fish were added or fish are being overfed) or it could be because something (temperature/pH/space) changed in the system that reduced the population of bacteria.

12. When will we be able to tell if our system start-up worked?

When the system is at equilibrium. We will know it worked and our system is ready for production when we feed the fish the same amount of feed each day and the ammonia and nitrite are at 0 ppm and the nitrates are not building or declining.

LESSON 6

Plants in Aquaponics Systems

Teacher's Guide

Time to Complete

3–4 CLASS PERIODS

PREPARATION

Prerequisite Knowledge

While not required, a quick review of plant anatomy may be beneficial for students.

Previous knowledge of osmosis is required.

Vocabulary

absorption
chlorophyll
chlorosis
flaccid
hypertonic
hypotonic
macronutrients
micronutrients
necrosis
nutrient deficiency
osmosis
photosynthesis
primary nutrient
reflection
turgid

Learning Objectives

- Students will identify and explain the parts of the plant cell involved in the process of photosynthesis.
- Students will use a model to explain how photosynthesis connects to the global carbon cycle.
- Students will obtain and communicate information about what types of plants are suitable for an aquaponics system.
- Students will obtain and communicate information about what nutrients plants need and compare them to what is provided by the aquaponics system.
- Students will observe plants to determine the symptoms of different nutrient deficiencies.
- Students will conduct investigations to determine and connect the symptoms of potassium deficiency to osmosis and turgor pressure.
- Students will conduct investigations to examine the plants in their system, identify nutrient deficiencies, and develop metrics for assessing the health of their plants.

NGSS Alignment

HS-LS1-5.	Use a model to illustrate how photosynthesis transforms light energy into stored chemical energy.
HS-LS2-2.	Use mathematical representations to support and revise explanations based on evidence about factors affecting biodiversity and populations in ecosystems of different scales.
HS-LS2-3.	Construct and revise an explanation based on evidence for the cycling of matter and flow of energy in aerobic and anaerobic conditions.
HS-LS2-5.	Develop a model to illustrate the role of photosynthesis and cellular respiration in the cycling of carbon among the biosphere, atmosphere, hydrosphere, and geosphere.
HS-LS2-4.	Use mathematical representations to support claims for the cycling of matter and flow of energy among organisms in an ecosystem.

Prep Work

- ☐ Print copies of the worksheet for students.
    - ☐ Worksheet: *Ready, Set, Grow!*
    - ☐ Answer key at the end of this lesson plan

- ☐ Print copies of the *Plant Nutrient Guide* for use during plant observations.
- ☐ Decide in advance if you want to use any of the optional activities detailed below during the discussion of photosynthesis. Each activity will require extra prep work.

Lesson Overview

1. Hand out the *Ready, Set, Grow!* worksheet at the beginning of class. Start class by having students brainstorm what plants to grow in your aquaponics system. After students have had a couple of minutes to list all the plants they would like to grow (slide 5), start the review.
2. The slide deck starts with a review of the plant cell, photosynthesis (optional review video on slide 11: https://www.youtube.com/watch?v=CMiPYHNNg28), and the global carbon cycle. Have students follow along as you review the plant cell (slides 7–10) and photosynthesis (starts on slide 11) by labeling the parts of the plant cell and writing the chemical equation on their worksheets. To make the discussion on photosynthesis more engaging, try one or more of the optional activities listed below.
3. The core of this section is focused on the nutrients needed and the symptoms plants show when they experience deficiencies. This is a rich and complex subject. The lesson only provides an introduction, so there is a great opportunity here for research projects on the biochemistry behind each nutrient deficiency.

   Finish the first part of the lesson after covering the potassium deficiency and turgor pressure (slide 48). Instruct students to complete the worksheet question on guard cells. If students are well-versed in transport mechanisms, ask them to hypothesize whether potassium is transported passively or actively.
4. Begin the second part of the lesson (slides 50–68) by completing the slide deck. By the end, students should begin to understand that certain plants will be more difficult than others to grow in the system. Have students reflect on which of their initial choices for plants would make the most sense based on the knowledge they now have.
5. Next, have the students as a group decide on their standards for "good plant observations." Students often make observations too quickly, so quickly that they miss important details—for example, the presence of aphids. Thus, it's best to have them create a set of expectations that they hold each other to.
6. Visit your system and have students take pictures of their plants for future comparison. Photos will help them notice any changes over time. Make sure someone tests or has recently tested the pH of the system. Once photos have been taken and pH tested, students should make plant observations using the *Plant Nutrient Guide* as a resource and recording observations on their worksheets. The plant nutrient guide has a table with nutrients and symptoms of deficiencies on the front and a graphic that illustrates nutrient uptake by plants at different pH values on the back. When used together, these materials can help students identify which nutrients plants may be lacking and if the deficiency is a result of pH levels.
7. Time permitting, reconvene and have students share their plant observations and discuss whether the system seems in good health. If your system has not been up and running long enough for deficiencies to appear, bring students back to the classroom and use the supplemental slides found at the end of the slide deck (slides 69–77) to make observations and determine what nutrient is missing from the system.

Optional Activities

What Type of Light Is Best?

Conduct the experiment proposed on slide 23. The cheapest way to do this is to buy color filters and fix them over plants underneath white light. Then, students can measure growth over time. Alternatively, if your students are up to a challenge, they could design an experiment to test the effect of different colored lights on flowering, fruiting, and foliage growth. If budget permits, blue and red grow lights would be best for this experiment.

Real Plant Deficiencies

We highly recommend raising a wide variety of plants in your aquaponics system, even if it is well-known that some will have nutrient deficiencies. This would create a perfect educational opportunity. Students will see firsthand that different plants require different levels and types of nutrients to flourish, and they can then apply the knowledge learned in the lecture to identify the deficiencies. To take this activity one step further, have students create their own class-made guide to nutrient deficiencies using pictures of their own system. This would be more valuable than the photos used in the lecture to illustrate deficiencies. For example, iron deficiency could be shown with strawberries and potassium deficiency with cucumber or tomatoes. Generally, try a few different fruiting vegetable plants because they will often show nutrient deficiencies.

Germination Experiment

Conduct experiments on the most efficient way to germinate seeds prior to entering the aquaponics system. Variables to test include presence/absence of light, distance to light source, presence/absence of nitrates, material (e.g., rockwool vs. peat), presence/absence of wind, and varying levels of humidity and temperature.

Water Quality Lab

The plants in your system have different requirements than your fish do. Have students collect water samples and test for primary plant nutrients (nitrogen, phosphorous, and potassium). If time and resources permit, include testing for calcium and magnesium in this mini lab.

Additional Resources

- *Illustrated Definitions of Plant Problems*: https://extension.purdue.edu/extmedia/ID/ID-319-W.pdf
- *Small-Scale Aquaponic Food Production: Integrated Fish and Farming* (pdf provided)
- AQU@TEACH OUTPUT 4: *Aquaponics Textbook for Higher Education*, chapter 5 (pdf provided)
- *Complete Guide for Growing Plants Hydroponically* (1st edition): https://www.researchgate.net/profile/Arvind-Singh-21/post/Hydroponics-for-Radish/attachment/5ed08b130294e50001c34553/AS%3A896381426360321%401590725394972/download/1.pdf
- *Hydroponic Salad Crop Production*: https://cropking.com/catalog/books/hydroponic-salad-crop-production
- *Hydroponic Food Production: A Definitive Guidebook for the Advanced Home Gardener and the Commercial Hydroponic Grower* (8th edition): https://www.amazon.com/dp/0367678756

Access the Lesson 6 slide deck by scanning the QR code below or following the link provided.

https://docs.lib.purdue.edu/aquaponics/6

Plants in Aquaponics Systems

Plant Nutrient Guide

Nutrient	*Deficiency Symptoms*
Nitrogen	• Fully yellow older leaves (nitrogen is mobile) • Slowed growth
Phosphorus	• Purple/brown spots on older leaves often along the veins (phosphorus is mobile) • More common in fruiting plants
Potassium	• Brown spots or yellowing edges in older leaves, called scorching (potassium is mobile) • Stunted growth • Wilted appearance
Calcium	• Splotchy burning on the edges of new growth (calcium is immobile) • Fairly common in aquaponics systems
Iron	• Yellow/pale leaves but green veins in new growth (iron is immobile) • Specific to certain plants
Magnesium	• Total leaf chlorosis in old leaves (magnesium is mobile) • Marginal necrosis (death) • Easily confused with nitrogen deficiencies

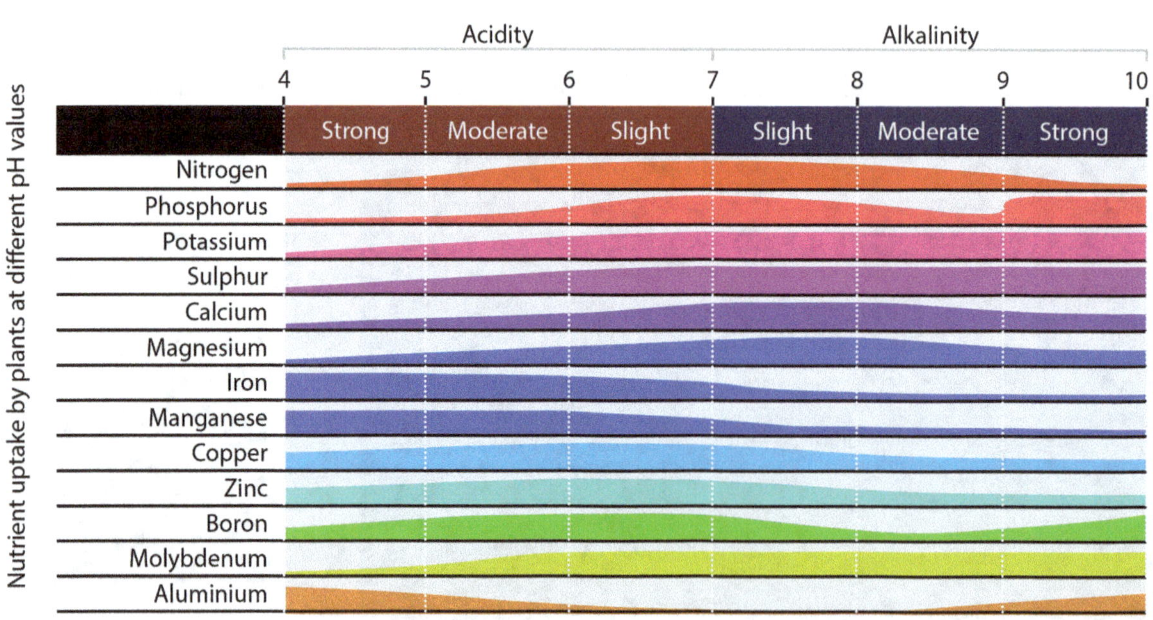

Nutrient uptake by plants at different pH values. Figure courtesy of Illinois–Indiana Sea Grant, adapted from RGJ Aquaponics.

LESSON 6

Plants in Aquaponics Systems

Worksheet

Ready, Set, Grow!

Name ______________________________ Date ___________ Class Period _______

1. What plants do you want to grow in your system?

Plant Cell Review

2. Label the parts of the plant cell by using words found in the word box. Write definitions for the cell wall, vacuole, and cytoplasm.

☐ Cell membrane ☐ Cell wall ☐ Chloroplast ☐ Chromosomes

☐ Cytoplasm ☐ Nucleus ☐ Vacuole

Name ________________________________ Date ___________ Class Period ______

3. Write the chemical equation for photosynthesis.

4. Potassium is a common deficiency in aquaponics systems. Potassium is, in part, responsible for the *turgor pressure* in plant cells, which is the mechanism responsible for opening and closing the stomata, the ports that regulate gas exchange for the plant.

   Using the model, describe the process of opening and closing the stomata with regard to the concentration of potassium. Speculate what would happen to the stomata if the plant was deficient in potassium. In your answer, use the words *osmosis*, *hypertonic*, *hypotonic*, *turgid*, and *flaccid* when describing this process.

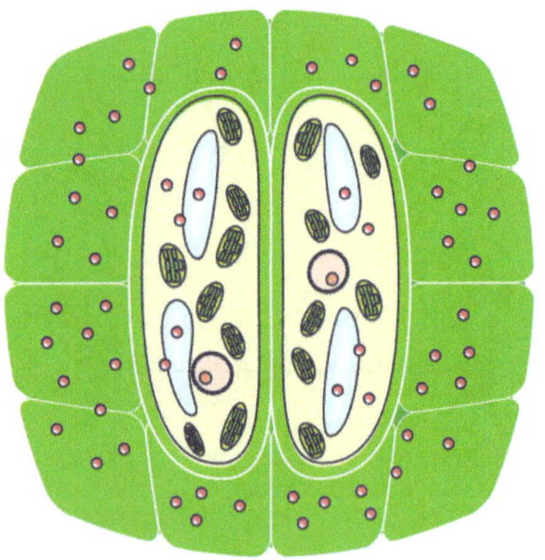

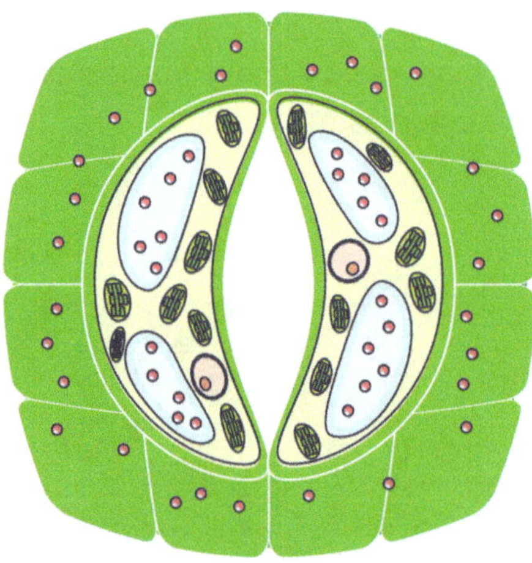

Name ______________________________ Date __________ Class Period ______

Plant Observations

5. It is important to catch problems early in an aquaponics system so you can make corrections quickly. This is only possible if you make thorough observations of your system. Start by making general observations of your plants, recording them in the space below. Is any particular type of plant doing better than others? Are certain areas doing better than others? Is there any discoloration? Are there pests? If you suspect any nutrient deficiencies, speculate on what the deficiency is and why.

6. Select one plant, record which one you've chosen, and make detailed observations and growth measurements (height, number of leaves, size of leaves). Construct a data table with your observations.

7. Test the pH of your water. Consult the chart on plant nutrient uptake at different pH levels. Make a claim about the uptake of any nutrient that you are concerned about. Use evidence and reasoning to support your concern.

8. Why is pH important to plants?

LESSON 6

Plants in Aquaponics Systems

Worksheet Answer Key

Ready, Set, Grow!

Name ______________________ Date __________ Class Period ______

1. What plants do you want to grow in your system?

Answers will vary. Answer should include the names of two or more plants.

Plant Cell Review

2. Label the parts of the plant cell by using words found in the word box. Write definitions for the cell wall, vacuole, and cytoplasm.

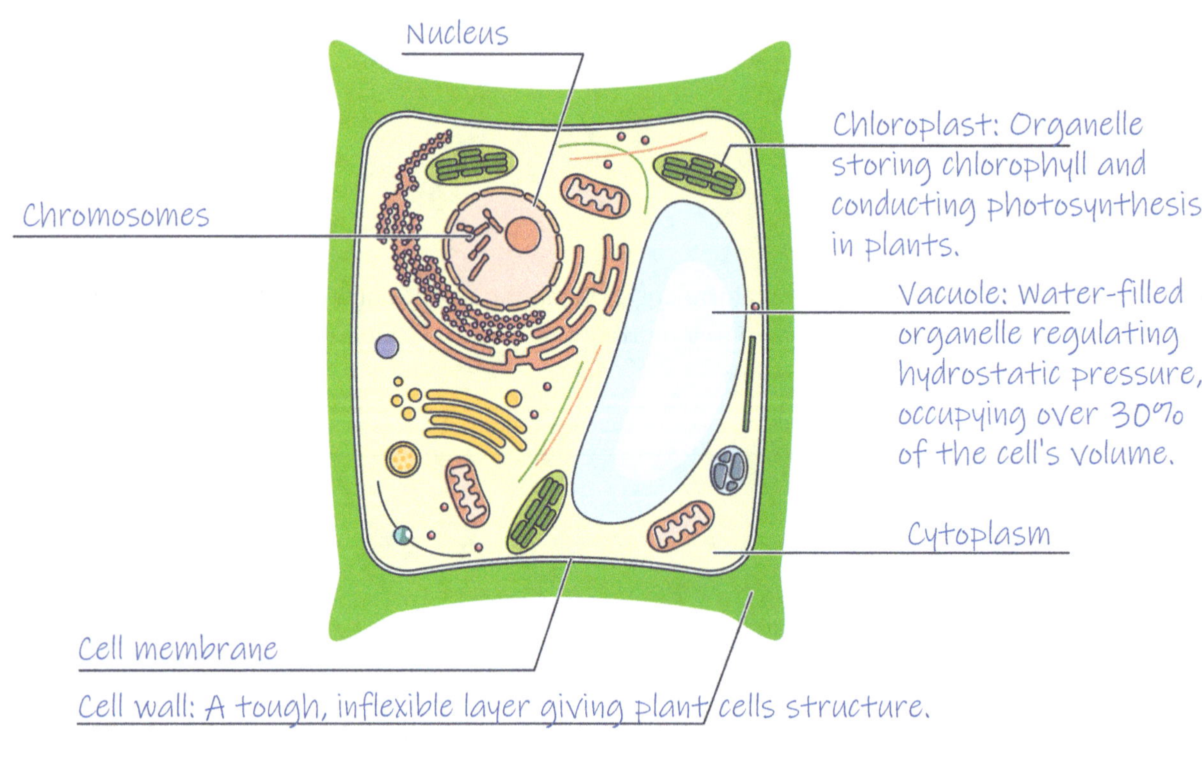

☐ Cell membrane	☐ Cell wall	☐ Chloroplast	☐ Chromosomes
☐ Cytoplasm	☐ Nucleus	☐ Vacuole	

Name ______________________________ Date __________ Class Period ______

3. Write the chemical equation for photosynthesis.

$6CO_2 + 6H_2O \leftrightharpoons C_6H_{12}O_6 + 6O_2$

4. Potassium is a common deficiency in aquaponics systems. Potassium is, in part, responsible for the *turgor pressure* in plant cells, which is the mechanism responsible for opening and closing the stomata, the ports that regulate gas exchange for the plant.

Using the model, describe the process of opening and closing the stomata with regard to the concentration of potassium. Speculate what would happen to the stomata if the plant was deficient in potassium. In your answer, use the words *osmosis*, *hypertonic*, *hypotonic*, *turgid*, and *flaccid* when describing this process.

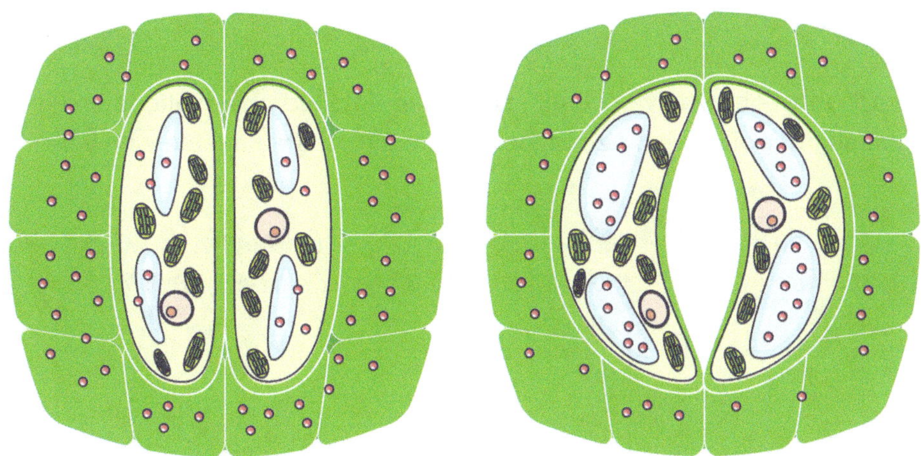

When the stoma is open, there is a high concentration of K^+ in the vacuoles, which draws water into the vacuoles because the solution outside is now hypotonic. The cells become turgid, which forces the stoma open.

When the stoma is closed, there is a low concentration of K^+ in the vacuoles, which draws water out because the solution outside of the vacuole is now hypertonic. The cells become flaccid and the stoma closes.

If the plant was potassium deficient, it wouldn't be able to drive up the concentration in the vacuoles and the cells would never become turgid. Thus, the stomata would never open and the plant would be limited in its ability to exchange gases.

Name ______________________________ Date ___________ Class Period ______

Plant Observations

5. It is important to catch problems early in an aquaponics system, so you can make corrections quickly. This is only possible if you make thorough observations of your system. Start by making general observations of your plants, recording them in the space below. Is any particular type of plant doing better than others? Are certain areas doing better than others? Is there any discoloration? Are there pests? If you suspect any nutrient deficiencies, speculate on what the deficiency is and why.

   Answers will vary based on student's observations.

6. Select one plant, record which one you've chosen, and make detailed observations and growth measurements (height, number of leaves, size of leaves). Construct a data table with your observations.

   Answers will vary based on the plant selected. Answers should identify what plant was selected, a description of what was observed (plant condition), and growth measurements.

7. Test the pH of your water. Consult the chart on plant nutrient uptake at different pH levels. Make a claim about the uptake of any nutrient that you are concerned about. Use evidence and reasoning to support your concern.

   Answers will vary based on pH test results. Answers should include a claim, evidence (pH level and information from the chart), and the reasoning to support the claim.

8. Why is pH important to plants?

   Water pH influences the availability of water nutrients to plants. If the pH level is high or low, even if the nutrients are present the plant will not be able to uptake it and the plants will be deprived of nutrients required for healthy growth.

LESSON 7

Fish in Aquaponics Systems

Teacher's Guide

Time to Complete

3 CLASS PERIODS

PREPARATION

Prerequisite Knowledge

Introductory knowledge on biological macromolecules. Proteins (particularly enzymes), lipids, carbohydrates, vitamins would be useful, but this lesson could serve as the introduction to these molecules.

Vocabulary

amino acids
carbohydrates
enzymes
fat-soluble
isozymes
kinetics
lipids
omega-3 and omega-6 fatty acids
poikilothermic
protein
water-soluble

Learning Objectives

- Students will obtain information about some common fish health issues encountered in aquaponics systems.
- Students will obtain information about how temperature and pH affect their fish.
- Students will obtain information about the optimum pH for all organisms in the aquaponics system and how to achieve the optimum pH.
- Students will create a model of a complete fish diet and compare this with their current fish food.
- Students will calculate the correct amount of food to feed their fish.
- Students will select fish for their aquaponics system.
- Students will observe and assess the health of their fish.

NGSS Alignment

HS-ETS1-3. Evaluate a solution to a complex real-world problem based on prioritized criteria and trade-offs that account for a range of constraints, including cost, safety, reliability, and aesthetics as well as possible social, cultural, and environmental impacts.

Prep Work

- ☐ Print copies of the *Buffer Mini Lab* worksheet for students.
- ☐ Set up lab stations for the Buffer Mini Lab. Each station should include a water sample from your system, a device to measure pH (pH pen or chemical test), titration supplies to measure alkalinity, baking powder, a scale or measuring device, vinegar, and a graduated cylinder or other vessel for measuring volume.
- ☐ Print copies of the worksheet for students.
    - ☐ Worksheet: *Fish Food Assessment*
    - ☐ Answer key at the end of this lesson plan
- ☐ Prepare printed copies or a way to display the ingredient list and description of the specific fish food used for your system.

- ☐ Print copies of *Fish Food Guide: What Do Your Fish Need for a Complete Diet?* for student use.
- ☐ Decide in advance if you want to use either of the optional activities detailed below on the final day of this lesson. Each activity will require extra prep work.

Lesson Overview

1. The slide deck begins with a quick overview of fish anatomy, which emphasizes some major anatomical differences between humans and fish. The focus then shifts to why water temperature and pH are so important to fish health and growth. These sections make connections to chemical kinetics and specialized enzymes, as well as to selecting organisms that all thrive at the same pH levels.
2. Next, the students will learn about what food to feed their fish, how much, and how often. The slide deck is light on details about the contents of the food, but this is the focus of (a) *Fish Food Guide: What Do Your Fish Need for a Complete Diet?* and the *Fish Food Assessment* worksheet, which can be used as an introduction or assessment for students' understanding of biological macromolecules, and (b) the *Selecting the Best Food for Your Fish* activity that can be found on slide 41.
3. The presentation finishes with the students brainstorming what factors are important in picking a fish species for their system, common fish selected for aquaponics, and what they should observe about their fish.
4. Prior to handing out the *Fish Food Assessment* worksheet, pass out printed copies of the ingredient list and description of the specific fish food used for your system, or display the information on a screen for students to see.
5. End the lesson with one or more optional activities (details below) and a short quiz (*Fish in Aquaponics*) to bring all of the concepts covered in this lesson together.

Suggested Schedule

Day 1	Fish anatomy, pH, and temperature (slides 3–32). End with *Buffer Mini Lab*.
Day 2	Feeding, selecting, and observing your fish (slides 33–46). Complete the *Fish Food Assessment* worksheet and/or the *Selecting the Best Food for Your Fish* research activity (slide 41).
Day 3	Fileting and cooking fish or Harvest Debate (Optional Activities). End with short *Fish in Aquaponics* quiz.

Additional Resources

- IISG's Students Ask Scientists: Video Chats: https://iiseagrant.org/education/students-ask-scientists

Buffer Mini Lab

Print the Buffer Mini Lab worksheet. To prep for the activity, set up lab stations. Each station should include a water sample from your system, a device to measure pH (pH pen or chemical test), titration supplies to measure alkalinity, baking powder, a scale or measuring device, vinegar, and a graduated cylinder or other vessel for measuring volume. Go through the instructions on the handout. After students have completed the activity, have them return to their seats to share their experience and discuss their results and conclusions. End with the question *How can we adjust pH and the carbonate buffer in the system?* (slide 32).

Optional Activities

Filet and Cook Fish

A big hurdle in increasing student interest in aquaponics is getting them to view their fish as a source of food. Many students don't frequently eat fish and generally seem to find fish gross. One of the most effective ways to break through this barrier is to actually cook and eat fish with your students. Despite the many reasons fish are nutritionally and environmentally beneficial, students' decisions about whether or not to eat fish usually come down to taste. So, if possible, we highly encourage teaching your students how to fillet and cook fish. Fish tacos or a fish fry are often well-received meals. Herbs and vegetables from your system can also be harvested for the meal. The act of harvesting and preparing their own meal, in theory, should be very empowering to your students. Try to engage them by appealing to their flair for drama and correctly point out that these are good survival skills for when the zombie apocalypse arrives.

- Teach your students how to fillet a fish: https://www.youtube.com/watch?v=sEue4x5FASw
- Interesting science on the taste of fish based on how they were killed. **WARNING: This video shows a fish being killed.** https://www.youtube.com/watch?v=TS4AM9mPX-8

Harvest Debate

Harvesting their fish is often a divisive subject among students. Holding a debate on whether or not to harvest their fish is also a great way to engage students about the merits of eating fish. Debate points could include the nutritional value of fish compared to other meats, the carbon footprint associated with raising fish, water usage, moral arguments of killing animals, and the taste. Split your class into two groups and assign one to be pro–fish harvesting and one to be anti–fish harvesting. Each group should develop four arguments for their side, and each group is allowed to make counterpoints after the other group's argument. Some good resources for your students to draw from are listed below.

- Health Benefits of Fish: https://www.doh.wa.gov/communityandenvironment/food/fish/healthbenefits
- The Carbon Footprint of Different Foods: https://www.visualcapitalist.com/visualising-the-greenhouse-gas-impact-of-each-food/
- Pescatarian Diets: https://www.healthline.com/nutrition/pescatarian-diet
- Water Usage: https://foodprint.org/issues/the-water-footprint-of-food/
- The Human Diet from an Evolutionary Perspective: https://www.nature.com/scitable/knowledge/library/evidence-for-meat-eating-by-early-humans-103874273/
- Evolution of the Human Diet: https://www.nationalgeographic.com/foodfeatures/evolution-of-diet/
- Diet and Chronic Health Conditions: https://www.greenfacts.org/en/diet-nutrition/index.htm

Additional Resources

- Know Your H2O testing kit (loanable test kit): https://iiseagrant.org/education/loanable-kits/
- *Know Your H2O Quick Reference Guide for Testing Water Quality*: https://iiseagrant.org/education/know-your-h2o/
- Recirculating aquaculture buffering (University of Tennessee): https://web.utk.edu/~rstrange/wfs556/html-content/04-buffering.html
- Eat Midwest Fish recipes: https://eatmidwestfish.org/recipe/%20
- Eat Midwest Fish cooking demonstration videos: https://eatmidwestfish.org/recipes/cooking-demos/

Access the Lesson 7 slide deck by scanning the QR code below or following the link provided.

https://docs.lib.purdue.edu/aquaponics/7

LESSON 7

Fish in Aquaponics Systems

Fish Food Guide: What Do Your Fish Need for a Complete Diet?

Fish in aquaculture or aquaponics rely entirely on you (the farmer) to provide a complete diet to promote health and growth. Selecting high-quality fish food is one of the most important decisions you'll make for the health of your aquaponics system. In this guide, you'll learn about the different components in the diet of your fish (which are actually not too different from your own) and apply that knowledge to design a model of the perfect food for your fish. You will then compare this to the fish food you are currently using and assess its quality. Along the way, you'll learn about the chemical structures and roles of the different biological macromolecules.

Proteins

Proteins are linked amino acids, of which there are about 20 commonly seen in nature. These 20 different amino acids can be combined in an infinite number of ways, creating a huge diversity of proteins with a myriad of unique jobs in organisms (for example, enzymes are proteins, which catalyze specific chemical reactions in the body). Fish can synthesize 10 of those amino acids given that they have the materials, but they cannot make the other 10 and thus these amino acids must be provided by their diet.

Ten required amino acids in fish food:

Methionine	Arginine	Threonine	Tryptophan	Histidine
Isoleucine	Lysine	Leucine	Valine	Phenylalanine

If the fish food does not provide a complete breakdown of the amino acid content, you may be able to assess the content based on the food source. Fish food made from plant protein, like soybean meal, is often low in methionine, whereas bacterial and yeast proteins are frequently low in methionine and lysine. Foods made from these sources will likely need to be supplemented by fishmeal, which provides both methionine and lysine.

The total amount of protein that is required varies between species, depending on whether the fish is herbivorous, omnivorous, or carnivorous. Also, to promote growth, smaller and younger fish usually require a higher percentage of protein than their larger and older counterparts. Typically, fish use protein for growth, but if the fish aren't getting enough carbohydrates or fats, proteins can also be used for energy. While high-protein diets are advantageous for increased growth, protein is the most expensive component of fish food, so a hallmark of cheap fish food is low protein content.

Species	*% Protein Required for an Average Adult*
Shrimp	30–35
Catfish	28–32
Tilapia	35–40
Hybrid striped bass	38–42
Trout	40–45

Lipids (Fats)

Lipids are high-energy molecules with about twice the energy density of proteins and carbohydrates. The high energy content comes from long carbon chains that get metabolized. Fats are metabolized slowly, so they are considered a "slow" source of energy. Fats are an important component of a complete diet, not just for their energy, but also because many minerals and vitamins are fat-soluble and require fat to be absorbed by the body. Lipids typically make up about 7% to 15% of fish diets.

Specific fats required:

Omega-3 fatty acids	EPA (eicosapentaenoic acid) Linolenic acid	0.5%–2.0% of dry diet Linolenic acid is often a required supplement for freshwater fish
Omega-6 fatty acids	Linoleic acid	Required by tilapia

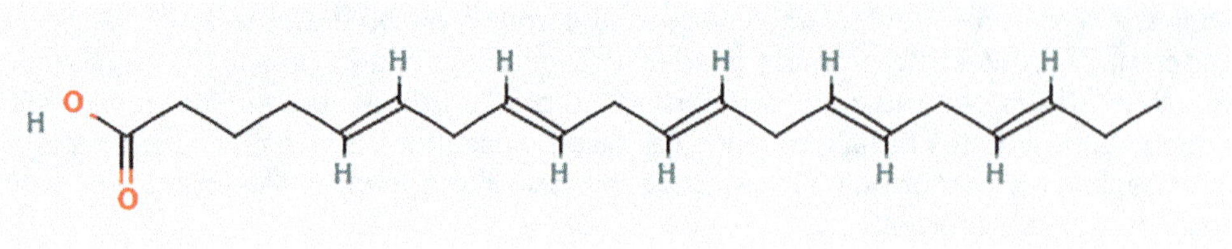

EPA, omega-3 fatty acid. From https://pubchem.ncbi.nlm.nih.gov/compound/5282847.

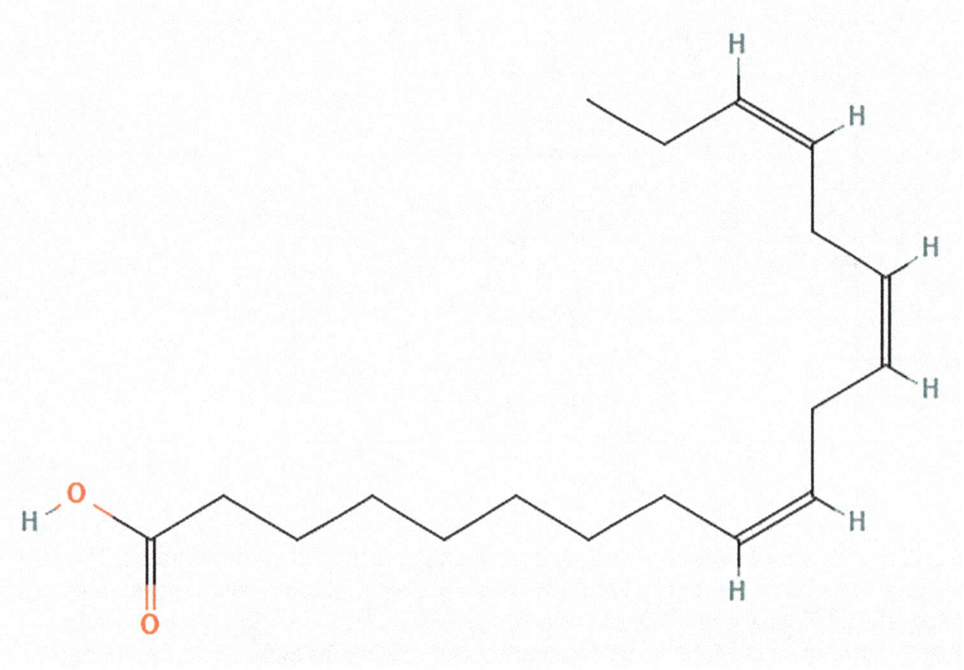

Linolenic acid, omega-3 fatty acid. From https://pubchem.ncbi.nlm.nih.gov/compound/5280934.

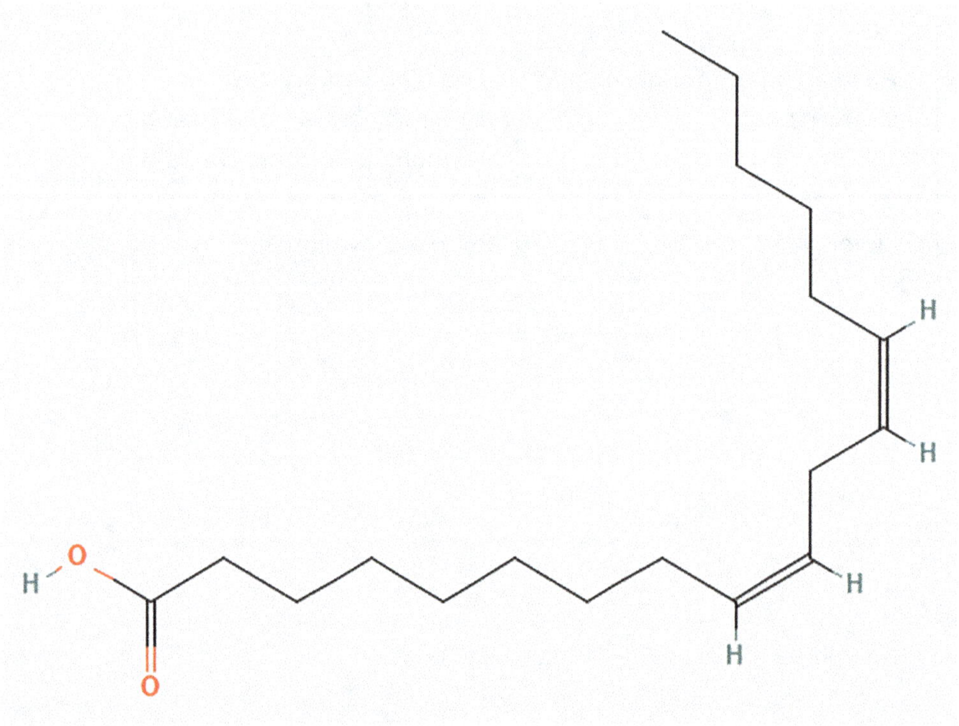

Linoleic acid, omega-6 fatty acid. From https://pubchem.ncbi.nlm.nih.gov/compound/5280450.

Carbohydrates

Often known as the fast-energy source, carbohydrates are also the cheapest source of energy in fish diets. Carbohydrates are technically not required in fish food but are often added as a filler to reduce the cost of manufacturing the food.

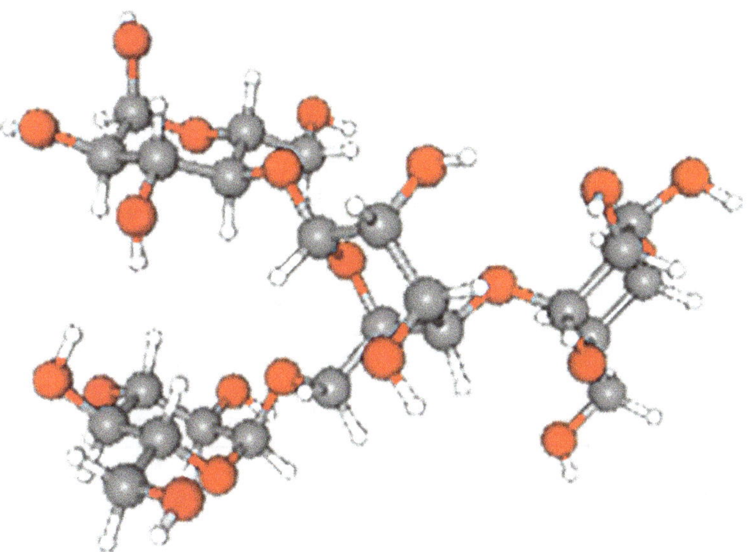

Glycogen, which is repeating units of glucose. From https://pubchem.ncbi.nlm.nih.gov/compound/439177#section=3D-Conformer.

Fish store carbohydrates such as glycogen, which is a huge network of linked glucose molecules. While this is a major source of energy for mammals (which of course includes humans), fish can't use glycogen as efficiently. Feeds for carnivorous fish are typically less than 20% carbohydrate, and for omnivorous fish they can reach up to 25% to 45% carbohydrate. While this is acceptable, make sure the carbohydrate content isn't so high that the food is deficient in proteins and fats.

Vitamins

Vitamins are an essential part of most animal diets (including ours), but most animals can't synthesize these complex molecules and must get them from food. Consequently, you should see the vitamins listed below in the ingredients on fish food packaging. As mentioned earlier, some vitamins are fat-soluble and others are water-soluble. This is an important distinction and plays a large role in whether the vitamin is absorbed by the body or allowed to pass through. Fish may show specific symptoms for each vitamin deficiency, but typically, you'll just notice reduced growth.

Water-Soluble Vitamins

B vitamins (thiamine, riboflavin, niacin, pyridoxine, biotin, folic acid, cobalamins)	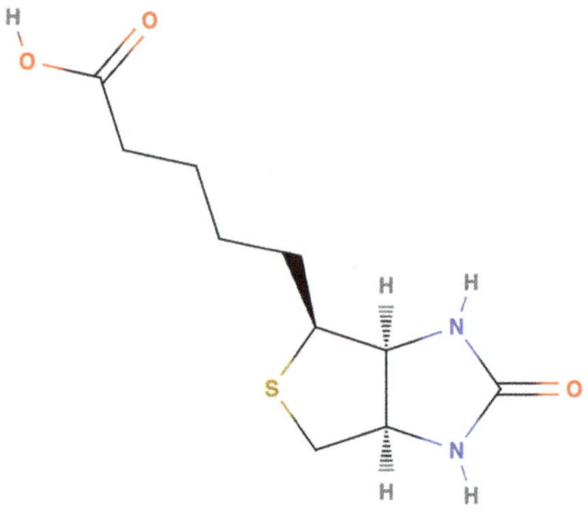 Biotin.
Inositol	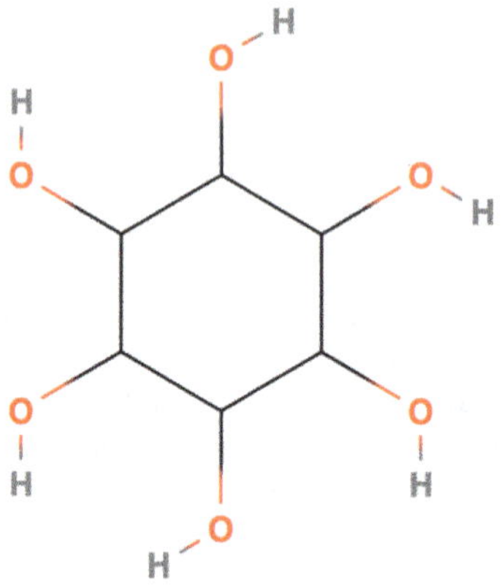
Vitamin C[1]	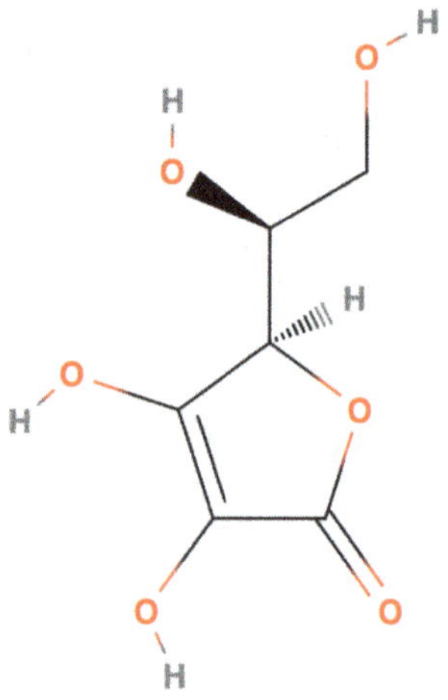

1. Especially important for enhancing the immune system of fish.

Fat-Soluble Vitamins

Vitamin A (retinol, beta-carotene)	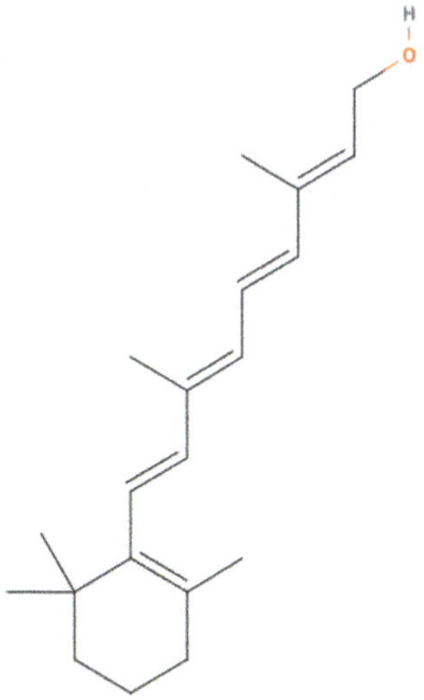 H O Retinol.
Vitamin D	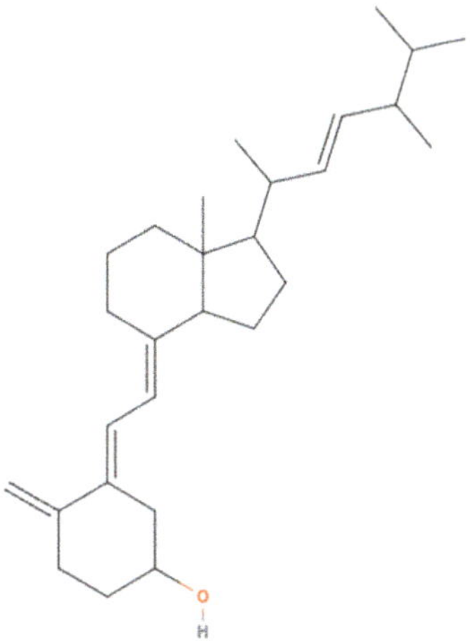 O H
Vitamin E	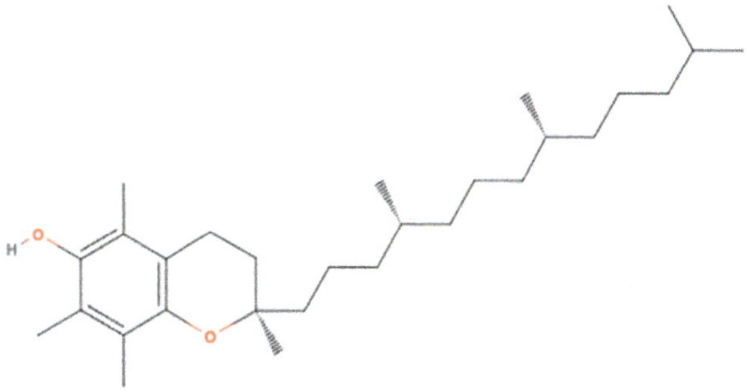 H O O
Vitamin K	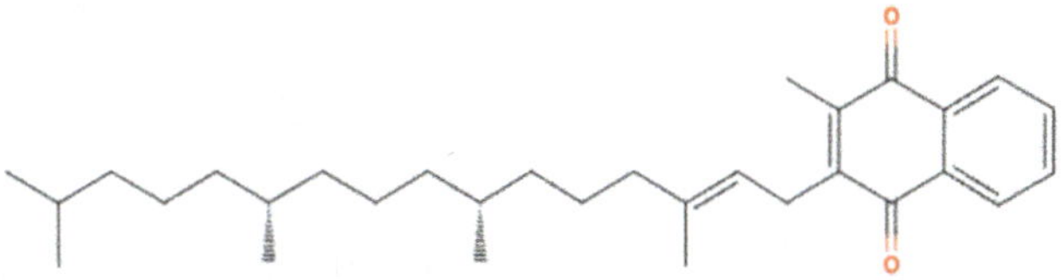 O O Vitamin K1.

Minerals

Minerals are the inorganic elements your fish need in small amounts to maintain osmotic balance and bone formation and to synthesize specific enzymes and hormones. These include calcium, sodium, chloride, potassium, chlorine, sulfur, phosphorus, magnesium, iron, copper, chromium, iodine, manganese, zinc, and selenium. You should recognize some of these as elements needed by your plants as well. While fish food can serve as a source of these nutrients for your plants, to minimize waste, the foods are often carefully tailored to meet only the needs of fish, so it is not wise to rely on fish food for plant nutrients.

Adapted from S. Craig, L. A. Helfrich, D. D. Kuhn, and M. H. Schwarz, *Understanding Fish Nutrition, Feeds, and Feeding* (Virginia Cooperative Extension Publication 420-256, 2017), https://vtechworks.lib.vt.edu/server/api/core/bitstreams/24c04f50-8d2f-4b2d-9f8a-9ec3684537a1/content.

LESSON 7

Fish in Aquaponics Systems

Worksheet

Buffer Mini Lab

Name ______________________________ **Date** __________ **Class Period** ______

Lab Group Members ____________________ ____________________

____________________ ____________________

____________________ ____________________

____________________ ____________________

____________________ ____________________

Introduction

When scientists do experiments, they often record their activities in a journal or lab notebook. Today, you are going to create a lab journal entry to keep a record of what you do. Include the names of all lab group members, the date, page numbers, starting pH and alkalinity levels, hypothesis, your procedures, results, and conclusion. Follow the instructions and use the section headers to guide you as you plan and complete your experiment.

Instructions

1. Record the names of lab group members and the date.
2. Establish benchmarks: Measure the pH and alkalinity of your water sample and record findings to establish what the pH and alkalinity levels are before anything is added.
3. Hypothesis: Select what you will add to your water sample (baking soda or vinegar) and come up with a hypothesis.
4. Procedures: Describe your procedures for testing your hypothesis. What steps will you take, and how will you measure the results?
5. Results: Record your results.
6. Conclusion: What do you conclude based on your findings? What can you add to your system to adjust pH and the carbonate buffer in the system?

Name ______________________________ Date __________ Class Period ______

Establish Benchmarks

Beginning pH:

Beginning alkalinity:

Hypothesis

(What do you think will happen when you add ______________________ to your water sample?)

Procedures

Results

Conclusion

LESSON 7

Fish in Aquaponics Systems

Worksheet

Fish Food Assessment

Name ________________________________ **Date** ___________ **Class Period** _______

Instructions: Answer the questions below.

1. Proteins are large structures made of linked amino acids. Assess the structures of the amino acids listed. Based on the similarity in their structures, where do you think amino acids create bonds to form long chains? (Hint: an -OH group will be replaced by -NH.)

2. Draw the peptide (small protein) of valine-valine-valine.

3. Proteins can be tens of thousands of amino acids long, which allows for HUGE chemical diversity in their structures. Calculate how many different peptides with a length of 10 amino acids you could make from the 20 most common amino acids in nature.

4. Compare the structures of the omega-3 and omega-6 fatty acids. What do the "3" and "6" mean to the type of structure?

Name ______________________________ Date __________ Class Period ______

5. Compare the structures of the water-soluble and fat-soluble vitamins. You'll notice that structurally they are quite diverse, but you should be able to identify a commonality between all the water-soluble vitamins and all the fat-soluble vitamins. What is it? Explain how this commonality relates to the vitamins being water- or fat-soluble. Remember, water is H_2O and fats have carbon-hydrogen structures.

6. After reading through the fish food assessment, design the perfect food for the fish in your system. Include the percentage of protein, the protein source, the percentage of fats, any specific fats needed, the percentage of carbohydrates, the vitamins needed, and the minerals needed.

7. Now compare the food that you designed with the description and ingredients listed on your fish food. Categorize each ingredient as a source of protein, carbohydrate, lipid, mineral, or vitamin (some ingredients may be added for other reasons—mark the ingredients you're not sure about). Assess the quality of your fish food. Make a claim about whether or not you endorse this fish food. Use evidence and reasoning to support your claim.

LESSON 7

Fish in Aquaponics Systems

Quiz

Fish in Aquaponics

Name ______________________________ **Date** __________ **Class Period** ______

Instructions: Answer the questions below.

Fish Anatomy

1. In one or two sentences, describe the function of the following parts of a fish.

    a. Gills:

    b. Swim bladder:

Temperature and pH

2. Why is it important to fish that temperature is controlled in an aquaponics system?

3. What is pH?

4. What is the optimum pH level in water for fish?

5. Why is pH important to fish?

6. What is the optimum level of pH in the water for nitrifying bacteria?

Name ______________________________ **Date** ___________ **Class Period** ______

7. Why is pH important to bacteria?

8. What is the optimum pH range for an aquaponics system, and why?

Bonus

9. What are four interrelated areas of science an aquaponics system addresses to be effective?

LESSON 7

Fish in Aquaponics Systems

Worksheet Answer Key

Buffer Mini Lab

Name ______________________________ **Date** __________ **Class Period** ______

Lab Group Members

_____________________ _____________________

_____________________ _____________________

_____________________ _____________________

_____________________ _____________________

_____________________ _____________________

Introduction

When scientists do experiments, they often record their activities in a journal or lab notebook. Today, you are going to create a lab journal entry to keep a record of what you do. Include the names of all lab group members, the date, page numbers, starting pH and alkalinity levels, hypothesis, your procedures, results, and conclusion. Follow the instructions and use the section headers to guide you as you plan and complete your experiment.

Instructions

1. Record the names of lab group members and the date.
2. Establish benchmarks: Measure the pH and alkalinity of your water sample and record findings to establish what the pH and alkalinity levels are before anything is added.
3. Hypothesis: Select what you will add to your water sample (baking soda or vinegar) and come up with a hypothesis.
4. Procedures: Describe your procedures for testing your hypothesis. What steps will you take, and how will you measure the results?
5. Results: Record your results.
6. Conclusion: What do you conclude based on your findings? What can you add to your system to adjust pH and the carbonate buffer in the system?

Name ______________________________ Date ___________ Class Period _______

Establish Benchmarks

Beginning pH: Answers will vary based on the water sample. Test the samples before class to determine pH.

Beginning alkalinity: Answers will vary based on the water sample. Test the samples before class to determine alkalinity (mg/L).

Hypothesis

(What do you think will happen when you add baking soda/vinegar to your water sample?)

If baking soda is added to water, the pH and alkalinity will increase.
If vinegar is added to water, the pH and alkalinity will decrease.

Procedures

Answers will vary from group to group. Here is one example:

1. Pour 200 mL of sample water into a 250 mL beaker.
2. Measure out 25 grams of baking soda.
3. Add 25 grams of baking soda to the water.
4. Mix until the baking soda is dissolved.
5. Measure the pH and alkalinity of your water sample and record findings.

Results

End pH: Answers will vary based on the experimental design, but as long as students add enough baking soda to change the pH, the pH will be higher than the beginning pH.

End alkalinity: Answers will vary based on the experimental design, but as long as students add enough baking soda to change the alkalinity, the alkalinity will be higher than the beginning alkalinity.

Conclusion

Answers will vary based on procedures. Here are a couple of examples:

Adding a base to water makes the water more basic; therefore, when the pH/alkalinity of the water is low, baking soda can be added to make the pH and alkalinity go up.

Adding an acid to water makes it more acidic; therefore, when the pH/alkalinity of the water is high, an acid can be added to make the pH and alkalinity go down.

LESSON 7

Fish in Aquaponics Systems

Worksheet Answer Key

Fish Food Assessment

Name ______________________________ **Date** __________ **Class Period** _______

Instructions: Answer the questions below.

1. Proteins are large structures made of linked amino acids. Assess the structures of the amino acids listed. Based on the similarity in their structures, where do you think amino acids create bonds to form long chains? (Hint: an -OH group will be replaced by -NH.)

   A peptide bond is formed between the carbonyl carbon and the NH group of another amino acid, which replaces the -OH group.

2. Draw the peptide (small protein) of valine-valine-valine.

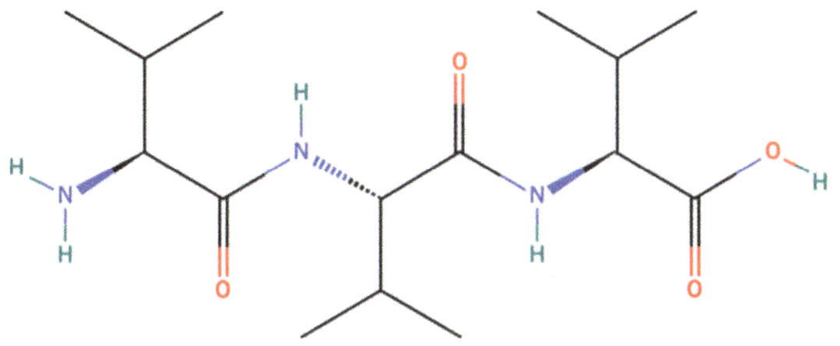

3. Proteins can be tens of thousands of amino acids long, which allows for HUGE chemical diversity in their structures. Calculate how many different peptides with a length of 10 amino acids you could make from the 20 most common amino acids in nature.

   $20^{10} = 20 \times 20 \times 20 \times 20 \times 20 \times 20 \times 20 \times 20 \times 20 \times 20 = 1.024 \times 10^{13}$ different combinations!

4. Compare the structures of the omega-3 and omega-6 fatty acids. What do the "3" and "6" mean to the type of structure?

   The 3 and 6 indicate where the first double bond is, as measured from the end of the fatty chain.

Name ________________________________ Date ___________ Class Period ______

5. Compare the structures of the water-soluble and fat-soluble vitamins. You'll notice that structurally they are quite diverse, but you should be able to identify a commonality between all the water-soluble vitamins and all the fat-soluble vitamins. What is it? Explain how this commonality relates to the vitamins being water- or fat-soluble. Remember, water is H_2O and fats have carbon-hydrogen structures.

All water-soluble vitamins have lots of -OH groups and, in some cases, NH groups. The fat-soluble vitamins are all carbon-hydrogen with no other types of atoms. -OH groups make vitamins water-soluble, whereas fat-soluble vitamins are only soluble in other carbon-hydrogen molecules. This is an example of "like-dissolves-like."

6. After reading through the fish food assessment, design the perfect food for the fish in your system. Include the percentage of protein, the protein source, the percentage of fats, any specific fats needed, the percentage of carbohydrates, the vitamins needed, and the minerals needed.

Answers will vary depending on the type of fish you have in your system. However, students should now be able to list and interpret the meaning of all the ingredients in their fish food.

7. Now compare the food that you designed with the description and ingredients listed on your fish food. Categorize each ingredient as a source of protein, carbohydrate, lipid, mineral, or vitamin (some ingredients may be added for other reasons—mark the ingredients you're not sure about). Assess the quality of your fish food. Make a claim about whether or not you endorse this fish food. Use evidence and reasoning to support your claim.

Answers will vary depending on the student-designed feed. Answers should include a comparison of ingredients and the claim, and evidence and reasoning to support the claim.

LESSON 7

Fish in Aquaponics Systems

Quiz Answer Key

Fish in Aquaponics

Name ________________________________ **Date** ___________ **Class Period** _______

Instructions: Answer the questions below.

Fish Anatomy

1. In one or two sentences, describe the function of the following parts of a fish.

   a. Gills: Fish gills absorb dissolved oxygen from the water and excrete waste products such as carbon dioxide and ammonia.

   b. Swim bladder: The swim bladder is a gas-filled organ that controls the buoyancy of the fish. It helps fish maintain the correct water depth.

Temperature and pH

2. Why is it important to fish that temperature is controlled in an aquaponics system?

   Most fish cannot maintain their own body temperature. They are cold-blooded, and their body temperature changes with the temperature of the water. If water is too warm, fish can die.

3. What is pH?

   pH is the measure of hydrogen ions, H^+, in solution.

4. What is the optimum pH level in water for fish?

   6.5–8.0

5. Why is pH important to fish?

   Fish are sensitive to the pH of their water. pH influences the chemistry of water. When the pH of the system water is high or low, elements and molecules in the water can be toxic to fish.

6. What is the optimum level of pH in the water for nitrifying bacteria?

   7.0–8.0

7. Why is pH important to bacteria?

Nitrifying bacteria thrive at different levels of pH in the system water. If the pH is too high or too low, the bacteria are not as active and will not be able to convert as much ammonia to nitrite or nitrite to nitrate as they could under optimal conditions.

8. What is the optimum pH range for an aquaponics system, and why?

Optimum pH for an aquaponics system is 6.8–7.2. When the system water is neutral or close to neutral, the fish, bacteria, and plants have good conditions in which to grow and thrive.

Bonus

9. What are four interrelated areas of science an aquaponics system addresses to be effective?

Along with plant sciences, aquaponics incorporates nitrifications, fish anatomy and nutrition, biology, and innovative high-tech agriculture.

LESSON 8

Our Modern Food System

Teacher's Guide

Time to Complete

45 MINUTES

PREPARATION

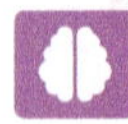

Prerequisite Knowledge

None

Vocabulary

colony collapse disorder
dead zones
food desert
food system
insecticides
monocultures
pesticides

Learning Objectives

- Students will construct an explanation about the complexity of society's current food system and create a model of the many steps that go into getting food on people's plates.
- Students will obtain and communicate information about some of the major drawbacks of our modern food system and agricultural practices.
- Students will construct an explanation about how aquaponics could be used to address these problems and act as a supplement to our modern food system.

NGSS Alignment

HS-ETS1-3. Evaluate a solution to a complex real-world problem based on prioritized criteria and trade-offs that account for a range of constraints, including cost, safety, reliability, and aesthetics as well as possible social, cultural, and environmental impacts.

Prep Work

- ☐ Print copies of the worksheets for students.
  - ☐ Worksheet: *Our Modern Food System*
  - ☐ Answer key at the end of this lesson plan
- ☐ Print copies of the jigsaw reading pages for use during plant observations. All students in each group should have a copy of the group's reading material.

Lesson Overview

1. Begin the lesson by playing the YouTube video *An Introduction to the Food System* (https://www.youtube.com/watch?v=Y3Yx2Mh3mSE). When the video is over, separate students into groups of four.
2. Each student should receive a copy of the *Our Modern Food System* worksheet and a copy of the jigsaw reading. Every student in each group will be responsible for reading a different passage and sharing their new knowledge with the group.
3. Have students do the jigsaw reading activity (see jigsaw reading procedures below for more details). Students should complete and summarize their reading. After groups share with their classmates what they learned, allow time for class discussion.

4. Play the YouTube video *Apples: From Farm to Table* (https://www.youtube.com/watch?v=66yQwR9OE4l) as an example of a food pathway before moving on to the worksheet.

5. Together, students should complete their worksheets, brainstorming how aquaponics can mitigate these problems. This lesson lends itself well to responsible action. As an optional activity, have students track their weekly food waste.

Jigsaw Reading Procedure

Jigsaw reading is a cooperative learning strategy that involves students working in groups to become experts on different parts of a text. The strategy breaks content into small chunks, requiring students in each group to become familiar with their assigned text and then share that knowledge with the rest of their group. The goal is that, by the end of the activity, all students are familiar with all the content.

Group 1	Group 2
Student A: Jigsaw Reading #1	Student E: Jigsaw Reading #1
Student B: Jigsaw Reading #2	Student F: Jigsaw Reading #2
Student C: Jigsaw Reading #3	Student G: Jigsaw Reading #3
Student D: Jigsaw Reading #4	Student H: Jigsaw Reading #4

For this jigsaw reading activity, there are four passages to read, so separate students into groups of four (sample group setup shown above). Hand out jigsaw reading assignments, one per student. Each member of the group should be given a different reading. Give students 3 to 5 minutes to read the assignment and complete the tasks. Next, give students in each group about 10 minutes to take turns sharing what they learned with each other. Finish the activity by discussing the readings as a class. Take this time to answer any questions that may have come up from the readings or during group discussions.

Video Links

An Introduction to the Food System: https://www.youtube.com/watch?v=Y3Yx2Mh3mSE

Apples: From Farm to Table: https://www.youtube.com/watch?v=66yQwR9OE4l

LESSON 8

Our Modern Food System

Jigsaw Reading #1

Name ________________________________ Date ___________ Class Period ______

Instructions: As you read the passage below, circle words you don't know and define them at the bottom. When you have finished reading, summarize what you read in one or two sentences.

Water Usage

While it may seem like we have plenty of water on earth (after all, the earth's surface is 71% water), only 3% of that is freshwater. And of that 3%, 83% is inaccessible because it is stored in glaciers. So, only 0.5% of the world's water is fresh *and* accessible.[1] This still amounts to a lot of water; however, the challenge is often in the redistribution of freshwater and assuring access to everyone. Currently, 2.7 billion people face water scarcity at least 1 month out of the year.[2]

On average, agricultural practices account for approximately 33% of our freshwater usage in America and 70% globally. Of that water, approximately 60% drains into the environment and requires purification before it can be used again. Consequently, it is of the utmost importance that agricultural practices begin using water more sustainably, especially as the world population continues to increase, and with it the demand for freshwater.

There are many ways to achieve this goal. First, we can adjust our methods. For example, because aquaponics and hydroponics recycle water, these techniques use approximately 10 times less water than traditional agricultural techniques.[3] Second, we can shift our diets away from water-intensive foods, particularly beef and pork products. For example, one hamburger requires 616 gallons of water to produce, most of which is used to grow the feed for the cattle. In contrast, 2 cups of broccoli only require 11 gallons of water. And if you don't want to give up meat in your diet, 1 pound of seafood can be produced with 1 pound of soybean product, whereas 1 pound of pork or beef requires 15 to 20 pounds of soybean product. That's a lot of water and resources for 1 pound of meat!

Summary Sentence

Definitions

Notes

1. "Water Facts—Worldwide Water Supply," Bureau of Reclamation, accessed February 28, 2021, https://www.usbr.gov/mp/arwec/water-facts-ww-water-sup.html.
2. "Water Scarcity," World Wildlife Fund, accessed February 28, 2021, https://www.worldwildlife.org/threats/water-scarcity.
3. "Hydroponics: A Better Way to Grow Food," National Park Service, accessed February 28, 2021, https://www.nps.gov/articles/hydroponics.htm.

LESSON 8

Our Modern Food System

Jigsaw Reading #2

Name ______________________________ Date __________ Class Period ______

Instructions: As you read the passage below, circle words you don't know and define them at the bottom. When you have finished reading, summarize what you read in one or two sentences.

Fertilizers

In the early 1800s, scientists discovered which elements were most essential to plant growth: nitrogen, phosphorus, and potassium. Later, fertilizer containing these elements was manufactured in the US and Europe. Now, many farmers use chemical fertilizers with nitrates and phosphates because they greatly increase crop yields.

However, fertilizers have come with another set of problems. The heavy reliance on chemicals has disturbed the environment, often destroying helpful species of animals along with harmful ones. Chemical use in agriculture may also pose a health hazard to people, especially through contaminated water supplies. High nitrite and nitrate concentrations are common to agricultural areas and are associated with a number of health issues, particularly for fetuses.

Nitrogen and phosphorus runoff from agricultural fields are some of the largest sources of pollution to coastal dead zones across the US. Perhaps the most infamous dead zone in the US is an 8,500 square mile area (about the size of New Jersey) in the Gulf of Mexico, not far from where the nutrient-laden Mississippi River drains into the gulf. Farms throughout the Midwest drain into the Mississippi and contribute to this nutrient overload. (In Indiana, for example, most of our state drains into the Wabash River, then the Ohio, and ultimately the Mississippi.) The gulf area dead zone has created massive challenges for the region's fishing industry that ripple throughout the economy. Low oxygen levels in the dead zone waters have led to reproductive problems for fish, which have resulted in low spawning rates and egg counts.

Summary Sentence

Definitions

This reading has direct excerpts from the following:

"The Art and Science of Agriculture," National Geographic Resource Library, accessed March 2, 2021, https://education.nationalgeographic.org/resource/the-art-and-science-of-agriculture/.

Clear Choices Clean Water Indiana, accessed March 2, 2021, https://indiana.clearchoicescleanwater.org/pledges/lawns/fertilizer-impacts/#void.

LESSON 8
Our Modern Food System
Jigsaw Reading #3

Name ______________________________ Date ___________ Class Period _______

Instructions: As you read the passage below, circle words you don't know and define them at the bottom. When you have finished reading, summarize what you read in one or two sentences.

Pesticides

Pesticides are chemicals or biological agents applied to crops to prevent damage caused by pests such as insects, weeds, or fungi. They have been incredibly important in maintaining high crop yields but have come with some disastrous environmental and health consequences.

Generally, pesticides are developed to damage, incapacitate, or kill an unwanted organism. Thus, it doesn't take much imagination to conclude that those chemicals could also be harmful to other beneficial organisms as well as to humans. In fact, many pesticides were developed from chemicals initially intended for chemical warfare. Over the years, scientists have discovered these drawbacks, which resulted in government regulation as well as continued research into effective but less toxic pesticides. However, it can be hard, if not impossible, to predict the long-term effects of a chemical in the environment.

For example, entomologists have recently observed a drop in bee populations, often described as colony collapse disorder. While there are multiple contributors to this problem, scientists have identified neonicotinoids, a class of insecticides, as a potential cause.[1] The loss of bee populations has significant consequences. Honeybees are the natural pollinator for many plants, including agricultural crops like almond trees and blueberries. Without honeybees, the production of these crops would plummet, or plants would have to be pollinated by hand. However, neonicotinoids were initially considered safe and heavily used throughout the 2000s. It wasn't until 2006 that scientists began making a connection between neonicotinoids and decreasing bee populations. Consequently, the US and European governments have started regulating their usage, but it is unclear how long-lasting their impact will be.

Summary Sentence

Definitions

Note

1. "Study Strengthens Link Between Neonicotinoids and Collapse of Honeybee Colonies," https://phys.org/news/2014-05-link-neonicotinoids-collapse-honey-bee.html.

LESSON 8

Our Modern Food System

Jigsaw Reading #4

Name ________________________________ Date __________ Class Period ______

Instructions: As you read the passage below, circle words you don't know and define them at the bottom. When you have finished reading, summarize what you read in one or two sentences.

Food Deserts, Health Concerns, and Monocultures

Food deserts are areas with limited or no access to fresh produce, a key component of a healthy diet.[1] Many chronic illnesses—such as obesity, diabetes, and cardiovascular diseases—are linked to diets high in sugar and unhealthy fats and low in fresh fruits and vegetables. However, in 2018, an estimated 200,000 people in the Indianapolis area were living in a food desert, as well as nearly 12 million people nationwide.[2] This may seem bizarre, especially in Indiana, given that when you drive across the state, the vast majority of land seems to be dedicated to agriculture.

However, most agriculture in the upper Midwest—like in Iowa, Illinois, and Indiana—produces crops not for direct consumption but rather for animal feed, processed foods, and biofuel. This results in acres of fields of either corn or soybeans, known as monocultures. Not only do these crops not provide much nutritional value to their communities, but they also can contribute to a number of environmental problems.

First, as you've learned, plants require specific nutrients to thrive. By growing the same crop over and over again, the soil is drained of the same nutrients year after year and often requires heavy application of fertilizers. Second, monocultures are extremely vulnerable to pests. Without crop diversity, a single pest can establish and destroy hundreds of square miles. Thus, monocultures encourage the use of pesticides, which come with their own drawbacks. All this said, the production of corn and soybeans is vital to the US economy and to providing food on a global scale. But, like everything, it is obviously not without its problems.

Summary Sentence

Definitions

Notes

1. "Food Access Research Atlas—Documentation," USDA Economic Research Service, updated January 5, 2025, accessed November 17, 2025, https://www.ers.usda.gov/data-products/food-access-research-atlas/documentation/; "What Are Food Deserts and How Do They Impact Health?," Medical News Today, accessed March 2, 2021, https://www.medicalnewstoday.com/articles/what-are-food-deserts#definition.
2. "Estimated 200,000 Indy Residents Live in Food Deserts," Savi, November 29, 2018, accessed March 2, 2021 https://www.savi.org/estimated-200000-indy-residents-live-in-food-deserts/#:~:text=Using%20recent%2C%20local%20data%20to,areas%2C%20and%20the%20Far%20Eastside; "11 Facts About Food Deserts," DoSomething, November 16, 2018, accessed March 2, 202,1 https://www.dosomething.org/us/facts/11-facts-about-food-deserts#:~:text=About%2023.5%20million%20people%20live,10%20miles%20from%20a%20supermarket.

LESSON 8

Our Modern Food System

Worksheet #1

Our Modern Food System

Name ______________________________ **Date** __________ **Class Period** ______

Introduction: Have you ever thought about how food gets to your plate? Every day, you (hopefully) eat three meals. Where did all that food come from? The process through which food is grown, cultivated, processed, and transported to you is collectively called a *food system*, and our food system in America is pretty complicated. Not many of us grow our own vegetables or raise our own livestock; instead, we rely on a complex global system to supply our food.

For example, think about eating an apple. That seems simple enough, right? Well, not exactly. That apple was grown, probably with fertilizers and pesticides (which required production and likely interrupted the natural ecosystem), then the apple was harvested by an employee and packaged. (Keep in mind that many agricultural employees are migrant workers who are underpaid and work without benefits like healthcare.) Then the apple was coated with wax to ensure that it maintained crispness as it was shipped across the country in an airplane or truck to your grocery store, where someone unpacked it and it was eventually sold to you. And that's not taking any processing into account. This all gets a lot more complicated if you want to make applesauce with that apple.

Instructions: Pick a food you've eaten in the last few days. Create a model of the steps that produced that food. Include as many steps as you can think of. If you've selected a meat, make sure to consider all the food production that went into creating the food for that animal.

The Food Supply Chain

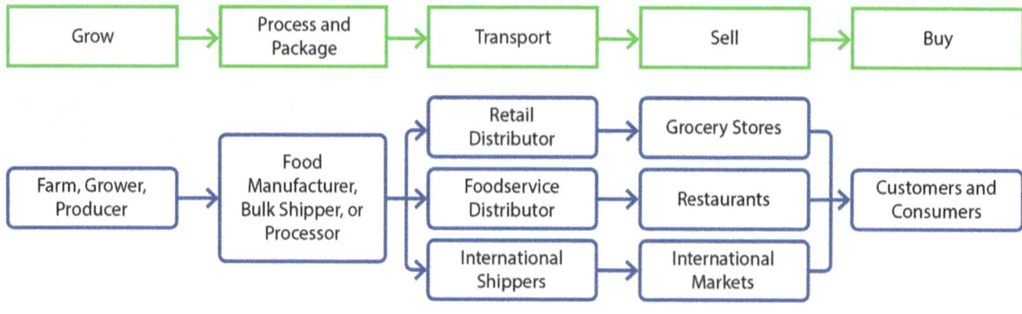

The Food Supply Chain: A Gallon of Milk

Figure courtesy of Illinois-Indiana Sea Grant. Adapted from the National Center for Agricultural Literacy.

Name ______________________________ Date __________ Class Period ______

Of course, in many ways, this process makes sense and is necessary. With a global population of 7.8 billion that continues to grow exponentially, and many people living in dense cities with little to no growing space, we rely on a global food system and modern agriculture. It's only been with modern inventions and technologies that feeding the entire world is even possible. For example, the Haber-Bosch process, the chemical reaction that produces ammonia for fertilizers, is estimated to be responsible for feeding half the world's population.[1] In other words, we wouldn't be able to feed approximately 4 billion people without that chemical innovation.

However, there are many drawbacks to modern agricultural practices. The primary concern of food politics over the last few decades has been to keep the cost of food down. Of course, this is an important consideration. If no one can afford the food, then what's the point of producing it? But the drive to keep food costs down comes with many negative repercussions. In the following readings, we will cover some of the major problems of modern agriculture and consider whether aquaponics or other technologies can offer solutions.

Note

1. "Haber-Bosch Process," BBC, accessed February 26, 2021, https://www.bbc.co.uk/programmes/p04f77rg.

LESSON 8

Our Modern Food System

Worksheet #2

Our Modern Food System

Name ______________________ **Date** __________ **Class Period** ______

1. In the table below, summarize the problem presented by modern agriculture for each issue. Then, brainstorm with your group on how aquaponics systems could reduce the impact of these problems.

Issue	*How Aquaponics Could Help*
Water usage:	
Fertilizers:	
Pesticides:	
Food deserts:	

Name ______________________________ Date __________ Class Period ______

2. What other advantages are offered by aquaponics?

3. Compared to traditional agriculture that often relies on irrigation, how is aquaponics more efficient?

4. What are some problems that aquaponics can't solve? Do you think aquaponics could replace traditional agriculture?

5. What are some major conclusions you can draw about the global food system?

6. How can aquaponics have a positive impact on urban and rural communities?

LESSON 8

Our Modern Food System

Jigsaw Reading #1 Answer Key

Name ______________________ Date __________ Class Period ______

Instructions: As you read the passage below, circle words you don't know and define them at the bottom. When you have finished reading, summarize what you read in one or two sentences.

Water Usage

While it may seem like we have plenty of water on earth (after all, the earth's surface is 71% water), only 3% of that is freshwater. And of that 3%, 83% is inaccessible because it is stored in glaciers. So, only 0.5% of the world's water is fresh *and* accessible.[1] This still amounts to a lot of water; however, the challenge is often in the redistribution of freshwater and assuring access to everyone. Currently, 2.7 billion people face water scarcity at least 1 month out of the year.[2]

On average, agricultural practices account for approximately 33% of our freshwater usage in America and 70% globally. Of that water, approximately 60% drains into the environment and requires purification before it can be used again. Consequently, it is of the utmost importance that agricultural practices begin using water more sustainably, especially as the world population continues to increase, and with it the demand for freshwater.

There are many ways to achieve this goal. First, we can adjust our methods. For example, because aquaponics and hydroponics recycle water, these techniques use approximately 10 times less water than traditional agricultural techniques.[3] Second, we can shift our diets away from water-intensive foods, particularly beef and pork products. For example, one hamburger requires 616 gallons of water to produce, most of which is used to grow the feed for the cattle. In contrast, 2 cups of broccoli only require 11 gallons of water. And if you don't want to give up meat in your diet, 1 pound of seafood can be produced with 1 pound of soybean product, whereas 1 pound of pork or beef requires 15 to 20 pounds of soybean product. That's a lot of water and resources for 1 pound of meat!

Summary Sentence

Agriculture competes for limited freshwater resources to produce food and other products. How food is grown and what foods we eat can both contribute to more sustainable water use.

Definitions

Glacier: an accumulation of snow and ice that slowly flows over land.

Notes

1. "Water Facts—Worldwide Water Supply," Bureau of Reclamation, accessed February 28, 2021, https://www.usbr.gov/mp/arwec/water-facts-ww-water-sup.html.
2. "Water Scarcity," World Wildlife Fund, accessed February 28, 2021, https://www.worldwildlife.org/threats/water-scarcity.
3. "Hydroponics: A Better Way to Grow Food," National Park Service, accessed February 28, 2021, https://www.nps.gov/articles/hydroponics.htm.

LESSON 8

Our Modern Food System

Jigsaw Reading #2 Answer Key

Name ______________________________ Date __________ Class Period ______

Instructions: As you read the passage below, circle words you don't know and define them at the bottom. When you have finished reading, summarize what you read in one or two sentences.

Fertilizers

In the early 1800s, scientists discovered which elements were most essential to plant growth: nitrogen, phosphorus, and potassium. Later, fertilizer containing these elements was manufactured in the US and Europe. Now, many farmers use chemical fertilizers with nitrates and phosphates because they greatly increase crop yields.

However, fertilizers have come with another set of problems. The heavy reliance on chemicals has disturbed the environment, often destroying helpful species of animals along with harmful ones. Chemical use in agriculture may also pose a health hazard to people, especially through contaminated water supplies. High nitrite and nitrate concentrations are common to agricultural areas and are associated with a number of health issues, particularly for fetuses.

Nitrogen and phosphorus runoff from agricultural fields are some of the largest sources of pollution to coastal dead zones across the US. Perhaps the most infamous dead zone in the US is an 8,500 square mile area (about the size of New Jersey) in the Gulf of Mexico, not far from where the nutrient-laden Mississippi River drains into the gulf. Farms throughout the Midwest drain into the Mississippi and contribute to this nutrient overload. (In Indiana, for example, most of our state drains into the Wabash River, then the Ohio, and ultimately the Mississippi.) The gulf area dead zone has created massive challenges for the region's fishing industry that ripple throughout the economy. Low oxygen levels in the dead zone waters have led to reproductive problems for fish, which have resulted in low spawning rates and egg counts.

Summary Sentence

Using chemical fertilizers benefits farmers by boosting production, but it can harm the environment and living things when excess chemicals get in the water table and in natural environments.

Definitions

Dead zone: area of water where the oxygen levels are too low for aquatic life to survive.

This reading has direct excerpts from the following:

"The Art and Science of Agriculture," National Geographic Resource Library, accessed March 2, 2021, https://education.nationalgeographic.org/resource/the-art-and-science-of-agriculture/.

Clear Choices Clean Water Indiana, accessed March 2, 2021, https://indiana.clearchoicescleanwater.org/pledges/lawns/fertilizer-impacts/#void.

LESSON 8

Our Modern Food System

Jigsaw Reading #3 Answer Key

Name________________________________ Date ___________ Class Period ______

Instructions: As you read the passage below, circle words you don't know and define them at the bottom. When you have finished reading, summarize what you read in one or two sentences.

Pesticides

Pesticides are chemicals or biological agents applied to crops to prevent damage caused by pests such as insects, weeds, or fungi. They have been incredibly important in maintaining high crop yields but have come with some disastrous environmental and health consequences.

Generally, pesticides are developed to damage, incapacitate, or kill an unwanted organism. Thus, it doesn't take much imagination to conclude that those chemicals could also be harmful to other beneficial organisms as well as to humans. In fact, many pesticides were developed from chemicals initially intended for chemical warfare. Over the years, scientists have discovered these drawbacks, which resulted in government regulation as well as continued research into effective but less toxic pesticides. However, it can be hard, if not impossible, to predict the long-term effects of a chemical in the environment.

For example, entomologists have recently observed a drop in bee populations, often described as colony collapse disorder. While there are multiple contributors to this problem, scientists have identified neonicotinoids, a class of insecticides, as a potential cause.[1] The loss of bee populations has significant consequences. Honeybees are the natural pollinator for many plants, including agricultural crops like almond trees and blueberries. Without honeybees, the production of these crops would plummet, or plants would have to be pollinated by hand. However, neonicotinoids were initially considered safe and heavily used throughout the 2000s. It wasn't until 2006 that scientists began making a connection between neonicotinoids and decreasing bee populations. Consequently, the US and European governments have started regulating their usage, but it is unclear how long-lasting their impact will be.

Summary Sentence

Pesticides used to control insects, weeds, and fungi in agricultural fields can be beneficial to crop production, but harmful to the environment and humans.

Definitions

Entomologist: a scientist who studies insects, spiders, and other arthropods.

Note

1. "Study Strengthens Link Between Neonicotinoids and Collapse of Honeybee Colonies," https://phys.org/news/2014-05-link-neonicotinoids-collapse-honey-bee.html.

LESSON 8

Our Modern Food System

Jigsaw Reading #4 Answer Key

Name ______________________ Date __________ Class Period ______

Instructions: As you read the passage below, circle words you don't know and define them at the bottom. When you have finished reading, summarize what you read in one or two sentences.

Food Deserts, Health Concerns, and Monocultures

Food deserts are areas with limited or no access to fresh produce, a key component of a healthy diet.[1] Many chronic illnesses—such as obesity, diabetes, and cardiovascular diseases—are linked to diets high in sugar and unhealthy fats and low in fresh fruits and vegetables. However, in 2018, an estimated 200,000 people in the Indianapolis area were living in a food desert, as well as nearly 12 million people nationwide.[2] This may seem bizarre, especially in Indiana, given that when you drive across the state, the vast majority of land seems to be dedicated to agriculture.

However, most agriculture in the upper Midwest—like in Iowa, Illinois, and Indiana—produces crops not for direct consumption but rather for animal feed, processed foods, and biofuel. This results in acres of fields of either corn or soybeans, known as monocultures. Not only do these crops not provide much nutritional value to their communities, but they also can contribute to a number of environmental problems.

First, as you've learned, plants require specific nutrients to thrive. By growing the same crop over and over again, the soil is drained of the same nutrients year after year and often requires heavy application of fertilizers. Second, monocultures are extremely vulnerable to pests. Without crop diversity, a single pest can establish and destroy hundreds of square miles. Thus, monocultures encourage the use of pesticides, which come with their own drawbacks. All this said, the production of corn and soybeans is vital to the US economy and to providing food on a global scale. But, like everything, it is obviously not without its problems.

Summary Sentence

Growing the same crop year after year has its own set of challenges that includes residents having limited access to healthy foods and more pollution caused by the increased use of pesticides.

Definitions

Monoculture: the cultivation of a single crop in a given area.

Notes

1. "Food Access Research Atlas—Documentation," USDA Economic Research Service, updated January 5, 2025, accessed November 17, 2025, https://www.ers.usda.gov/data-products/food-access-research-atlas/documentation/; "What Are Food Deserts and How Do They Impact Health?," Medical News Today, accessed March 2, 2021, https://www.medicalnewstoday.com/articles/what-are-food-deserts#definition.
2. "Estimated 200,000 Indy Residents Live in Food Deserts," Savi, November 29, 2018, accessed March 2, 2021 https://www.savi.org/estimated-200000-indy-residents-live-in-food-deserts/#:~:text=Using%20recent%2C%20local%20data%20to,areas%2C%20and%20the%20Far%20Eastside; "11 Facts About Food Deserts," DoSomething, November 16, 2018, accessed March 2, 202,1 https://www.dosomething.org/us/facts/11-facts-about-food-deserts#:~:text=About%2023.5%20million%20people%20live,10%20miles%20from%20a%20supermarket.

LESSON 8
Our Modern Food System
Worksheet #1 Answer Key

Our Modern Food System

Name ______________________________ **Date** __________ **Class Period** ______

Introduction: Have you ever thought about how food gets to your plate? Every day, you (hopefully) eat three meals. Where did all that food come from? The process through which food is grown, cultivated, processed, and transported to you is collectively called a *food system*, and our food system in America is pretty complicated. Not many of us grow our own vegetables or raise our own livestock; instead, we rely on a complex global system to supply our food.

For example, think about eating an apple. That seems simple enough, right? Well, not exactly. That apple was grown, probably with fertilizers and pesticides (which required production and likely interrupted the natural ecosystem), then the apple was harvested by an employee and packaged. (Keep in mind that many agricultural employees are migrant workers who are underpaid and work without benefits like healthcare.) Then the apple was coated with wax to ensure that it maintained crispness as it was shipped across the country in an airplane or truck to your grocery store, where someone unpacked it and it was eventually sold to you. And that's not taking any processing into account. This all gets a lot more complicated if you want to make applesauce with that apple.

Instructions: Pick a food you've eaten in the last few days. Create a model of the steps that produced that food. Include as many steps as you can think of. If you've selected a meat, make sure to consider all the food production that went into creating the food for that animal.

The Food Supply Chain

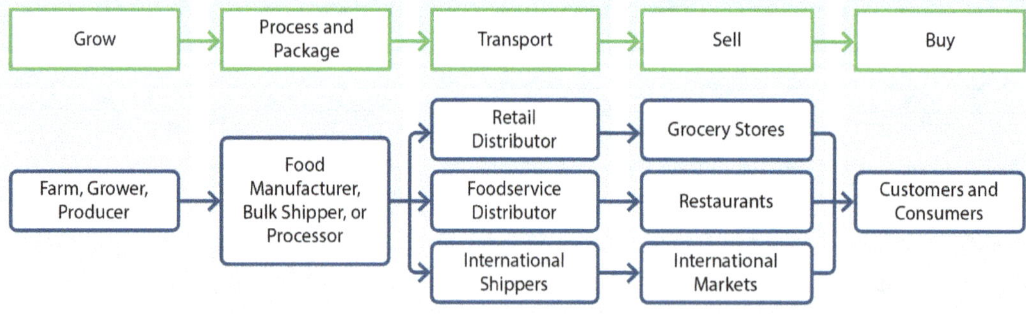

The Food Supply Chain: A Gallon of Milk

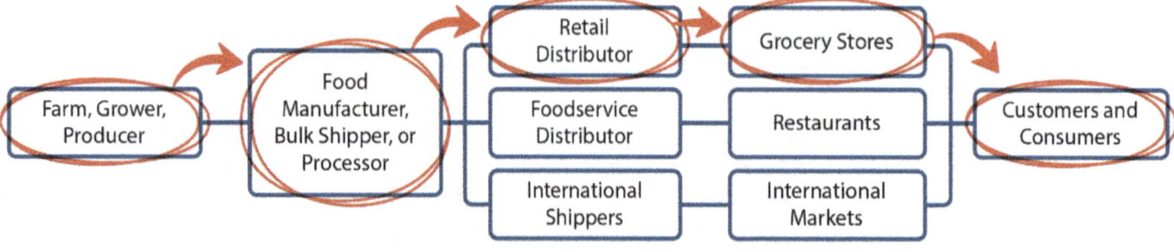

Figure courtesy of Illinois-Indiana Sea Grant. Adapted from the National Center for Agricultural Literacy.

Name ______________________________ Date __________ Class Period ______

Answers will vary. Answers should include one food and supply chain step. The supply chain should start with the farm and end with the buy/consumer.

Of course, in many ways, this process makes sense and is necessary. With a global population of 7.8 billion that continues to grow exponentially, and many people living in dense cities with little to no growing space, we rely on a global food system and modern agriculture. It's only been with modern inventions and technologies that feeding the entire world is even possible. For example, the Haber-Bosch process, the chemical reaction that produces ammonia for fertilizers, is estimated to be responsible for feeding half the world's population.[1] In other words, we wouldn't be able to feed approximately 4 billion people without that chemical innovation.

However, there are many drawbacks to modern agricultural practices. The primary concern of food politics over the last few decades has been to keep the cost of food down. Of course, this is an important consideration. If no one can afford the food, then what's the point of producing it? But the drive to keep food costs down comes with many negative repercussions. In the following readings, we will cover some of the major problems of modern agriculture and consider whether aquaponics or other technologies can offer solutions.

Note

1. "Haber-Bosch Process," BBC, accessed February 26, 2021, https://www.bbc.co.uk/programmes/p04f77rg.

LESSON 8

Our Modern Food System

Worksheet #2 Answer Key

Our Modern Food System

Name ______________________ **Date** __________ **Class Period** ______

1. In the table below, summarize the problem presented by modern agriculture for each issue. Then, brainstorm with your group on how aquaponics systems could reduce the impact of these problems.

Issue	*How Aquaponics Could Help*
Water usage: There is a limited amount of fresh water available on Earth. Traditional agriculture does not use water sustainably because 60% of water becomes runoff.	Aquaponics recycles the water used and uses approximately 10 times less water than traditional agriculture. Fish are also produced, which provides a healthy source of meat for a low water cost.
Fertilizers: To increase crop yield, fertilizers heavy in nitrogen and phosphorus are applied. This runs into water supplies and creates dead zones, which lower fish populations.	Aquaponics uses fish waste as fertilizer, reducing the overall need for manufactured fertilizer. Furthermore, aquaponics farming is separate from the environment, so it doesn't directly pollute water supplies.
Pesticides: To increase crop yield, pesticides are applied. However, these chemicals often have unintended and unknown effects on beneficial organisms such as honeybees.	Because most aquaponics farming is conducted indoors, there are generally fewer pests, which reduces the need for pesticides.
Food deserts: Poor diets are the root cause of many chronic health disorders. Despite being in the middle of agricultural fields, a lot of people in the Midwest live in food deserts with little or no access to fresh produce.	Local aquaponics businesses could provide much-needed access to fresh produce in areas designated as food deserts. Because they are run indoors, aquaponics farms can provide fresh produce in urban areas throughout the year.

Name ______________________ Date __________ Class Period ______

2. What other advantages are offered by aquaponics?

   - Can be conducted year-round if done indoors.
   - Good educational tool and community resource.
   - Small systems can easily be kept at home, providing fresh herbs and produce year-round.

3. Compared to traditional agriculture that often relies on irrigation, how is aquaponics more efficient?

   Aquaponics can significantly reduce daily freshwater requirements because of the efficient water cycles.

4. What are some problems that aquaponics can't solve? Do you think aquaponics could replace traditional agriculture?

   Aquaponics could not scale profitably to feed the world. Aquaponics comes with its own challenges and requires a lot of energy, especially in cold climates. It is best used as a supplemental system to traditional agriculture.

5. What are some major conclusions you can draw about the global food system?

   When it comes to our food systems, we have a lot of considerations to balance as a society: the cost of production, nutritional value, water usage, and environmental impacts. This is all on top of simply trying to produce enough food to keep up with the growing global population. It should be clear to the students that every decision we make to alter the environment has more consequences than we usually intend or anticipate.

6. How can aquaponics have a positive impact on urban and rural communities?

   Aquaponics offers local and fresh food. People can buy high-quality fish and vegetables in their own community, which reduces transportation costs and other barriers.

LESSON 9

Troubleshooting Water Quality Challenges

Teacher's Guide

Time to Complete

2–3 CLASS PERIODS

(Depending on the assessment type.)

PREPARATION

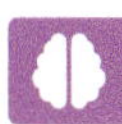

Prerequisite Knowledge

Safe water quality levels for the system

Procedures for testing common water quality parameters (DO, pH, temperature, ammonia, nitrate, nitrite)

Vocabulary

aphid

Learning Objectives

→ Students will conduct investigations and obtain information through research to generate a class guidebook for fixing common problems encountered in an aquaponics system.

NGSS Alignment

HS-ETS1-3. Evaluate a solution to a complex real-world problem based on prioritized criteria and trade-offs that account for a range of constraints, including cost, safety, reliability, and aesthetics as well as possible social, cultural, and environmental impacts.

Prep Work

- ☐ Print copies of the worksheet for students.
  - ☐ Worksheet: *Aquaponics Troubleshooting* (10 separate scenarios)
  - ☐ Answer key at the end of this lesson plan
- ☐ Decide in advance if you want to require students to create short presentations to share with the class (optional assignment details below). This may require extra prep work.

Lesson Overview

1. Hand out the *Aquaponics Troubleshooting* scenarios. Each student will receive 1 of 10 scenarios in which the aquaponics system is not thriving.
2. Students will use the curriculum's resources, as well as online research, to diagnose the problem, request further information, and propose solutions. The students should act as if they have all possible resources to fix their system. When compiled, the information from students' troubleshooting worksheets will create a guidebook for how to fix common problems in an aquaponics system.
3. Students can be instructed to do a peer review of each other's troubleshooting worksheets. Then, they can work together as a class to compile the best parts of each other's troubleshooting worksheets to create an ultimate master guide that can be shared with other science classes. Sharing the guidebook with other classes will allow more students to help care for the aquaponics system.

4. As these scenarios or other problems are encountered in the system, a new guidebook page should be added. Include pictures of the problem, how long the problem persisted, and how it was resolved.

Additional Information

The difficulty of the scenarios varies significantly. Scenarios 2, 3, 4, and 6 require research outside of the curriculum as well as more complex problem-solving. Scenarios 1, 5, 7, 8, 9, and 10 are relatively straightforward and can be answered using the material provided in the slide decks. Most problems have a "quick fix" solution. For example, in Scenario 1, the high nitrate concentration can be solved by a water change. However, students should also provide a long-term solution and consider the underlying cause. In this case, the plant-to-fish ratio is off and needs to be adjusted.

Optional Assignment

In addition to completing the worksheet, this project offers a good opportunity for short student presentations. Student groups would make a few slides and present the problem and solution(s) to their classmates. The class would then discuss the proposed solutions and any other possible underlying causes.

LESSON 9

Troubleshooting Water Quality Challenges

Worksheet

Aquaponics Troubleshooting: Scenario 1

Name ______________________ **Date** __________ **Class Period** _______

Instructions: Read the scenario below in which the system is not thriving. Use what you have learned in class to determine if you have enough information to propose what the issue is and come up with potential short- and long-term solutions. If not, do online research to collect the additional information required to answer the questions.

Scenario 1

You test the water nitrogen levels, and these are your results:

Ammonia: 0 ppm
Nitrite: 0 ppm
Nitrate: 120 ppm

1. Proposed issue(s):

2. Further information/testing/data needed:

3. Proposed solution(s):

LESSON 9

Troubleshooting Water Quality Challenges

Worksheet

Aquaponics Troubleshooting: Scenario 2

Name ________________________________ **Date** ___________ **Class Period** _______

Instructions: Read the scenario below in which the system is not thriving. Use what you have learned in class to determine if you have enough information to propose what the issue is and come up with potential short- and long-term solutions. If not, do online research to collect the additional information required to answer the questions.

Scenario 2

Older tomato plant leaves are yellowing, edges are scorched, and the plant appears limp. You test the water nitrogen levels, and these are your results:

Ammonia: 0 ppm
Nitrite: 0 ppm
Nitrate: 15 ppm

1. Proposed issue(s):

2. Further information/testing/data needed:

3. Proposed solution(s):

LESSON 9

Troubleshooting Water Quality Challenges

Worksheet

Aquaponics Troubleshooting: Scenario 3

Name ______________________________ **Date** __________ **Class Period** ______

Instructions: Read the scenario below in which the system is not thriving. Use what you have learned in class to determine if you have enough information to propose what the issue is and come up with potential short- and long-term solutions. If not, do online research to collect the additional information required to answer the questions.

Scenario 3

Algae is all over your fish tank and biofilter.

1. Proposed issue(s):

2. Further information/testing/data needed:

3. Proposed solution(s):

LESSON 9

Troubleshooting Water Quality Challenges

Worksheet

Aquaponics Troubleshooting: Scenario 4

Name ______________________________ **Date** __________ **Class Period** ______

Instructions: Read the scenario below in which the system is not thriving. Use what you have learned in class to determine if you have enough information to propose what the issue is and come up with potential short- and long-term solutions. If not, do online research to collect the additional information required to answer the questions.

Scenario 4

You find a dead fish in your tank.

1. Proposed issue(s):

2. Further information/testing/data needed:

3. Proposed solution(s):

LESSON 9

Troubleshooting Water Quality Challenges

Worksheet

Aquaponics Troubleshooting: Scenario 5

Name ______________________________ **Date** __________ **Class Period** ______

Instructions: Read the scenario below in which the system is not thriving. Use what you have learned in class to determine if you have enough information to propose what the issue is and come up with potential short- and long-term solutions. If not, do online research to collect the additional information required to answer the questions.

Scenario 5

You test the water nitrogen levels, and these are your results:

Ammonia: 0 ppm
Nitrite: 2 ppm
Nitrate: 5 ppm

A week ago, ammonia and nitrite were 0 ppm and nitrate was 15 ppm. However, since it's the beginning of winter, you note that the temperature in school is noticeably colder.

1. Proposed issue(s):

2. Further information/testing/data needed:

3. Proposed solution(s):

LESSON 9

Troubleshooting Water Quality Challenges

Worksheet

Aquaponics Troubleshooting: Scenario 6

Name ______________________________ **Date** __________ **Class Period** ______

Instructions: Read the scenario below in which the system is not thriving. Use what you have learned in class to determine if you have enough information to propose what the issue is and come up with potential short- and long-term solutions. If not, do online research to collect the additional information required to answer the questions.

Scenario 6

You notice aphids on some of your plants.

1. Proposed issue(s): (First, what are aphids?)

2. Further information/testing/data needed:

3. Proposed solution(s):

LESSON 9

Troubleshooting Water Quality Challenges

Worksheet

Aquaponics Troubleshooting: Scenario 7

Name ______________________________ **Date** ___________ **Class Period** _______

Instructions: Read the scenario below in which the system is not thriving. Use what you have learned in class to determine if you have enough information to propose what the issue is and come up with potential short- and long-term solutions. If not, do online research to collect the additional information required to answer the questions.

Scenario 7

You notice paleness and yellowing in new growth. You test your nitrogen levels, and these are your results:

Ammonia: 0 ppm
Nitrite: 0 ppm
Nitrate: 10 ppm

1. Proposed issue(s):

2. Further information/testing/data needed:

3. Proposed solution(s):

LESSON 9

Troubleshooting Water Quality Challenges

Worksheet

Aquaponics Troubleshooting: Scenario 8

Name ______________________________ **Date** __________ **Class Period** ______

Instructions: Read the scenario below in which the system is not thriving. Use what you have learned in class to determine if you have enough information to propose what the issue is and come up with potential short- and long-term solutions. If not, do online research to collect the additional information required to answer the questions.

Scenario 8

You notice full yellowing in older plant leaves. You test your nitrogen levels, and these are your results:

Ammonia: 0 ppm
Nitrite: 0 ppm
Nitrate: 0 ppm

1. Proposed issue(s):

2. Further information/testing/data needed:

3. Proposed solution(s):

LESSON 9

Troubleshooting Water Quality Challenges

Worksheet

Aquaponics Troubleshooting: Scenario 9

Name ________________________________ **Date** ___________ **Class Period** _______

Instructions: Read the scenario below in which the system is not thriving. Use what you have learned in class to determine if you have enough information to propose what the issue is and come up with potential short- and long-term solutions. If not, do online research to collect the additional information required to answer the questions.

Scenario 9

You notice brown discoloration on your fish as well as frayed fins. You haven't checked the nitrogen levels in over a week.

1. Proposed issue(s):

2. Further information/testing/data needed:

3. Proposed solution(s):

LESSON 9

Troubleshooting Water Quality Challenges

Worksheet

Aquaponics Troubleshooting: Scenario 10

Name ______________________________ **Date** __________ **Class Period** ______

Instructions: Read the scenario below in which the system is not thriving. Use what you have learned in class to determine if you have enough information to propose what the issue is and come up with potential short- and long-term solutions. If not, do online research to collect the additional information required to answer the questions.

Scenario 10

Your fish are gulping air at the surface of your tank.

1. Proposed issue(s):

2. Further information/testing/data needed:

3. Proposed solution(s):

LESSON 9

Troubleshooting Water Quality Challenges

Worksheet Answer Key

Aquaponics Troubleshooting: Scenario 1

Name ______________________________ **Date** __________ **Class Period** ______

Instructions: Read the scenario below in which the system is not thriving. Use what you have learned in class to determine if you have enough information to propose what the issue is and come up with potential short- and long-term solutions. If not, do online research to collect the additional information required to answer the questions.

Scenario 1

You test the water nitrogen levels, and these are your results:

Ammonia: 0 ppm
Nitrite: 0 ppm
Nitrate: 120 ppm

1. Proposed issue(s):

   The nitrate levels are too high in the system. The recommended level is approximately 20–50 ppm. The ratio of fish to plants is likely off in the system. Either there are too many fish (or they're being fed too much) or there are not enough plants. The ammonia and nitrite levels indicate that there are healthy levels of bacteria in the system.

2. Further information/testing/data needed:

   No further testing is required.

3. Proposed solution(s):

   Short-term solution: Do a partial water change (safely) to temporarily reduce the nitrate concentration.

   Long-term solution: If more plants can be added to the system, add plants. If this is not an option, reduce the number of fish, or if advisable, the amount of food the fish are receiving.

LESSON 9

Troubleshooting Water Quality Challenges

Worksheet

Aquaponics Troubleshooting: Scenario 2

Name ______________________________ **Date** ___________ **Class Period** _______

Instructions: Read the scenario below in which the system is not thriving. Use what you have learned in class to determine if you have enough information to propose what the issue is and come up with potential short- and long-term solutions. If not, do online research to collect the additional information required to answer the questions.

Scenario 2

Older tomato plant leaves are yellowing, edges are scorched, and the plant appears limp. You test the water nitrogen levels, and these are your results:

Ammonia: 0 ppm
Nitrite: 0 ppm
Nitrate: 15 ppm

1. Proposed issue(s):

   Potassium deficiency.

2. Further information/testing/data needed:

   A potassium level test should be completed. While the plant description matches potassium deficiency (and the nitrate levels are fine), and this is common in tomato plants, confirmation of low potassium levels in the system would be useful.

   It may also be useful to test calcium levels. If calcium is high, it can reduce potassium uptake.

   Lastly, a pH test would be useful to confirm that the plants can absorb potassium efficiently.

3. Proposed solution(s):

   If the tests come back and point to potassium deficiency, spray the leaves with a potassium chloride solution or add potassium hydroxide to the water.

LESSON 9

Troubleshooting Water Quality Challenges

Worksheet

Aquaponics Troubleshooting: Scenario 3

Name ______________________________ **Date** __________ **Class Period** ______

Instructions: Read the scenario below in which the system is not thriving. Use what you have learned in class to determine if you have enough information to propose what the issue is and come up with potential short- and long-term solutions. If not, do online research to collect the additional information required to answer the questions.

Scenario 3

Algae is all over your fish tank and biofilter.

1. Proposed issue(s):

   There is too much algae in the system, which could cause oxygen depletion or swings in the pH.

2. Further information/testing/data needed:

   Need to know if phosphorus was recently added to the system (high phosphorus concentrations result in algae growth).

3. Proposed solution(s):

   Clean the tank and reduce the light where algae is growing (cover the fish tank with a tarp to block light—this won't disrupt your fish or bacteria) and wait to add phosphorus until algae is under control. If the problem is not resolved by reducing the light, consider adding mechanical filtration to the system.

LESSON 9

Troubleshooting Water Quality Challenges

Worksheet

Aquaponics Troubleshooting: Scenario 4

Name ______________________________ **Date** __________ **Class Period** ______

Instructions: Read the scenario below in which the system is not thriving. Use what you have learned in class to determine if you have enough information to propose what the issue is and come up with potential short- and long-term solutions. If not, do online research to collect the additional information required to answer the questions.

Scenario 4

You find a dead fish in your tank.

1. Proposed issue(s):

   One or more water quality parameters could be out of range, creating an unsafe environment for the fish. Another possibility may be illness or injury.

2. Further information/testing/data needed:

   Remove the fish from the tank ASAP and examine the body for any signs of disease or exterior damage from other fish (check whether your fish are known to be cannibals at high stocking densities).

   Observe your other fish for lethargic behavior.

   Check the ammonia, nitrite, pH, and oxygen levels.

3. Proposed solution(s):

   The fish may have died from natural causes, in which case no changes are needed. Reduce the stocking density if death appeared to be related to aggressive or cannibalistic behavior.

   If ammonia/nitrite levels are high, reduce the levels immediately with a safe partial water change. Reduce feeding and add more surface area to increase biofiltration. If oxygen levels are low, consider adding another airstone or check for algal growth. If the pH is off, you may need to move your fish to a quarantine tank while your system equilibrates or you make chemical changes.

LESSON 9

Troubleshooting Water Quality Challenges

Worksheet

Aquaponics Troubleshooting: Scenario 5

Name ______________________________ **Date** __________ **Class Period** ______

Instructions: Read the scenario below in which the system is not thriving. Use what you have learned in class to determine if you have enough information to propose what the issue is and come up with potential short- and long-term solutions. If not, do online research to collect the additional information required to answer the questions.

Scenario 5

You test the water nitrogen levels, and these are your results:

Ammonia: 0 ppm
Nitrite: 2 ppm
Nitrate: 5 ppm

A week ago, ammonia and nitrite were 0 ppm and nitrate was 15 ppm. However, since it's the beginning of winter, you note that the temperature in school is noticeably colder.

1. Proposed issue(s):

   Nitrite levels are too high for the system. Fish are in danger. The *Nitrobacter* population or their activity likely fell with the cold temperatures.

2. Further information/testing/data needed:

   Check the pH, dissolved oxygen, and water temperature and research environmental conditions that result in decreased bacteria activity.

3. Proposed solution(s):

   Immediately reduce the concentration of nitrite with a safe partial water change (may need to quarantine your fish).

   Increase the temperature of the room if possible.

   Add more surface area to your system to increase the population of *Nitrobacter*, or immediately increase the population by adding a solution of bacteria.

LESSON 9

Troubleshooting Water Quality Challenges

Worksheet

Aquaponics Troubleshooting: Scenario 6

Name ______________________ **Date** __________ **Class Period** _______

Instructions: Read the scenario below in which the system is not thriving. Use what you have learned in class to determine if you have enough information to propose what the issue is and come up with potential short- and long-term solutions. If not, do online research to collect the additional information required to answer the questions.

Scenario 6

You notice aphids on some of your plants.

1. Proposed issue(s): (First, what are aphids?)

   Aphids are sap-sucking insects that are common pests in aquaponics systems. They collect on the stems of plants and are typically small and yellow.

   Aphids are harmful to plants. They suck the juices from leaves and stems, damaging plant parts. Aphid infestation can lead to stunted growth, attract ants, and provide food for fungus.

2. Further information/testing/data needed:

   No testing is required. Research safe solutions for getting rid of aphids.

3. Proposed solution(s):

   Solutions should be natural. Avoid the use of pesticides because the lack of pesticide use is a huge benefit of aquaponics systems. A concentrated garlic spray can be made and used to repel the aphids. If this doesn't work, ladybugs—which are natural predators of aphids—can be purchased and released on the system. This is a good example of biological control.

LESSON 9

Troubleshooting Water Quality Challenges

Worksheet

Aquaponics Troubleshooting: Scenario 7

Name ______________________________ **Date** __________ **Class Period** ______

Instructions: Read the scenario below in which the system is not thriving. Use what you have learned in class to determine if you have enough information to propose what the issue is and come up with potential short- and long-term solutions. If not, do online research to collect the additional information required to answer the questions.

Scenario 7

You notice paleness and yellowing in new growth. You test your nitrogen levels, and these are your results:

Ammonia: 0 ppm
Nitrite: 0 ppm
Nitrate: 10 ppm

1. Proposed issue(s):

   The plant description suggests an iron deficiency.

2. Further information/testing/data needed:

   It would be helpful to know what type of plants are affected to see if they are commonly iron deficient.

   Measure the iron levels in the system.

   Test the pH of the system to make sure that iron can be absorbed at that pH level.

3. Proposed solution(s):

   If the further testing indicates an iron deficiency, add iron chelate in the form of Fe-DTPA. You'll need to carefully calculate the amount of iron to add to the system.

LESSON 9

Troubleshooting Water Quality Challenges

Worksheet

Aquaponics Troubleshooting: Scenario 8

Name ______________________________ **Date** __________ **Class Period** ______

Instructions: Read the scenario below in which the system is not thriving. Use what you have learned in class to determine if you have enough information to propose what the issue is and come up with potential short- and long-term solutions. If not, do online research to collect the additional information required to answer the questions.

Scenario 8

You notice full yellowing in older plant leaves. You test your nitrogen levels, and these are your results:

Ammonia: 0 ppm
Nitrite: 0 ppm
Nitrate: 0 ppm

1. Proposed issue(s):

   Plants are showing nitrogen deficiency.

2. Further information/testing/data needed:

   You need to know why there is a nitrogen deficiency now. Were more plants recently added? Were the number of fish reduced?

   Test pH to make sure it's within a normal range for nitrogen absorption.

3. Proposed solution(s):

   If safe, add more fish or increase the amount of fish food. If this isn't a viable option, reduce the number of plants in the system. If you prefer to do neither, you may need to add more nitrogen to the system (although, this sort of defeats the purpose of aquaponics).

   If the pH is so far off as to interrupt nitrogen absorption, there are likely many other problems, and the pH needs to be adjusted ASAP.

LESSON 9

Troubleshooting Water Quality Challenges

Worksheet

Aquaponics Troubleshooting: Scenario 9

Name ________________________________ **Date** ___________ **Class Period** _______

Instructions: Read the scenario below in which the system is not thriving. Use what you have learned in class to determine if you have enough information to propose what the issue is and come up with potential short- and long-term solutions. If not, do online research to collect the additional information required to answer the questions.

Scenario 9

You notice brown discoloration on your fish as well as frayed fins. You haven't checked the nitrogen levels in over a week.

1. Proposed issue(s):

   The description of the fish suggests ammonia burns.

2. Further information/testing/data needed:

   Test nitrogen levels ASAP.

   Were there any changes made recently to the system? Why might the ammonia levels be high?

3. Proposed solution(s):

   If the ammonia is high in the system, a safe partial water change should be done immediately to drop the concentration of ammonia. May need to quarantine fish and treat with antibiotic solution.

   The long-term solution is to pinpoint why the ammonia is high and fix the underlying problem.

   - If more fish were added recently, increase the biofilter by adding more surface area for bacteria to grow on.
   - Was there a drop in temperature or pH that may have affected the *Nitrosomonas*? Readjust the temperature or pH. To expedite the process, you could add more *Nitrosomonas* to try to rectify the problem immediately.

LESSON 9

Troubleshooting Water Quality Challenges

Worksheet

Aquaponics Troubleshooting: Scenario 10

Name ______________________________ Date ___________ Class Period _______

Instructions: Read the scenario below in which the system is not thriving. Use what you have learned in class to determine if you have enough information to propose what the issue is and come up with potential short- and long-term solutions. If not, do online research to collect the additional information required to answer the questions.

Scenario 10

Your fish are gulping air at the surface of your tank.

1. Proposed issue(s):

   The fish behavior suggests a low dissolved oxygen content.

2. Further information/testing/data needed:

   Need to know the dissolved oxygen content. If it's below 4 mg/L, the oxygen content should be increased.

   Has anything changed recently in the system? Have the fish been overfed? Is algae growing? Is the temperature increased? Are there more fish?

3. Proposed solution(s):

   Short-term solution: Add an additional airstone to the system.

   Long-term solution: Determine why the dissolved oxygen is low. If something changed in the system to warrant a decrease in oxygen (e.g., an increase of fish), then an additional airstone can be the long-term solution. However, if there was no change in the system, the other issues need to be investigated.

LESSON 10

Building an Aquaponics Business

Teacher's Guide

Time to Complete

4–5 CLASS PERIODS

(Depending on the assessment style.)

PREPARATION

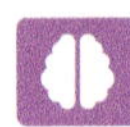

Prerequisite Knowledge

None

Vocabulary

None

Learning Objectives

- Students will synthesize all they have learned throughout the curriculum to develop a model of an aquaponics business plan.
- For their model, students will select plants and fish that meet their business goals and are scientifically compatible.
- Students will estimate the cost and revenue of their aquaponics business.

NGSS Alignment

HS-ETS1-3. Evaluate a solution to a complex real-world problem based on prioritized criteria and trade-offs that account for a range of constraints, including cost, safety, reliability, and aesthetics as well as possible social, cultural, and environmental impacts.

Prep Work

- ☐ Provide copies of the worksheet *Aquaponics Business Challenge* for students.
- ☐ Provide copies of the spreadsheet guides for students, either electronically or printed.
  - ☐ *Common Plants and Fish in Aquaponics*
  - ☐ *Aquaponics Cost and Revenue Analysis*
- ☐ Decide in advance which assessment style choice(s) you will offer students. Each style may require extra prep work.

Lesson Overview

1. Using the *Common Plants and Fish in Aquaponics* spreadsheet and the *Aquaponics Cost and Revenue Analysis* spreadsheet, students will work together in groups to build an aquaponics business plan. This project provides a great opportunity for students to flex their creativity as well as their business savvy. Encourage students to be as detailed as possible in their business designs.

   Assessment Options

   a. Worksheet: Students will complete the *Aquaponics Business Challenge* assignment sheet.

b. PowerPoint presentation: Students will present their business plan to the class. Bring in other teachers and administrators to act as judges to elevate the stakes of the assignment.

c. Shark Tank presentation: For a fun twist, set up the format of presentation using a model like Shark Tank so that students pitch their ideas to possible investors. Bring in other teachers and administrators and/or local business leaders to play the role of investors.

2. There is a short slide deck to introduce the project that includes a 2-minute video on vertical farming (https://www.youtube.com/watch?v=LCne-2BxkzM). The video features UP Vertical Farms (https://upverticalfarms.com/), a farm in British Columbia that uses controlled environment farming to grow leafy greens hydroponically. The video reinforces the science students have learned over the course of the curriculum (minus the fish component) and provides an example of a beautiful alternative to traditional agriculture.

3. Next, students should split into groups of two or three and begin planning their business. The spreadsheet guides provide approximate data for a handful of common plants and fish raised in aquaponics. If your students would like to research other options (e.g., different fish, cheaper source of food, more efficient light fixtures), encourage them to do so. However, for students who may be less enthused about additional research, the guides provide enough information for them to complete the assignment.

Additional Information

To estimate the cost and income for the fish and plants in their system, students will need to calculate the number of fish and plants they can harvest per year for their selections. For example, if a group chose to grow only basil in a large system, they would first need to calculate how many basil plants they could grow in that year:

$$\text{Total basil plants per year} = 1{,}500 \times 365/50 = {\sim}10{,}950$$

Thus, the total cost would be:

$$\text{Total cost of basil seeds per year} = 10{,}950 \times \$0.02 = \$219$$

And the total income would be:

$$\text{Total income of basil per year} = 10{,}950 \times \$7 = \$76{,}650$$

These are rough estimates, of course, and there are many unconsidered factors. For example, students should point out that the cost of the water itself, extra nutrients, and additional nutrient tests are not included. The cost of renting/owning the space is also not considered, nor how much they would pay themselves and their employees. It's also unclear whether there would be enough local demand for basil to justify growing so much. In other words, students could suggest doing some market research as part of their business proposal. There will be some holes in the final business plans, but the goal is to get students excited about aquaponics and provide a creative outlet to use the knowledge they've learned throughout the curriculum.

Access the Lesson 10 slide deck by scanning the QR code below or following the link provided.

https://docs.lib.purdue.edu/aquaponics/8

LESSON 10

Building an Aquaponics Business

Worksheet

Aquaponics Business Challenge

Name ______________________________ **Date** __________ **Class Period** ______

Business Name: ________________________________

1. **Proposal:** Imagine you are starting an aquaponics business, and you require investors to get your business started. Use the space below to write a proposal to persuade them to invest in your company. Start by determining the purpose and goals of your aquaponics system. Is it for profit? Is it for educational purposes? What community will it serve, and why would it be important for that community? Where will it be built? How will the size of the system as well as the fish and plants in your system meet these goals? Your selected fish and plants must be compatible; include the information necessary to confirm that your system is viable (see the spreadsheet *Common Plants and Fish in Aquaponics*). Include the projected revenue for your system. If it is negative, justify why your system should be subsidized by the community.

Name ______________________________ Date ___________ Class Period ______

2. **Economic plan:** Using the information provided in the two spreadsheets (*Common Plants and Fish in Aquaponics* and *Aquaponics Cost and Revenue Analysis*), use the sections below to calculate the start-up costs, annual operating costs, and annual income for your proposed business. If you have decided to grow plants or raise fish that are not on these spreadsheets, complete the research necessary to calculate their cost and income.

3. **Start-up cost:** What size system did you select?

4. **Energy cost:** Show your calculation for the annual energy cost for your selected system. Enter this value into your *Aquaponics Cost and Revenue Analysis* spreadsheet.

5. **Plant cost:** Show your calculation for the annual cost of plants in your system. Enter this value into your *Aquaponics Cost and Revenue Analysis* spreadsheet.

6. **Plant income:** Show your calculation for the annual income from the plants in your system. Make sure to account for the growing time and number of harvests. Enter this value into your *Aquaponics Cost and Revenue Analysis* spreadsheet.

7. **Fish cost:** Show your calculation for the annual cost of the fish in your system. You will need to calculate how many of your selected fish can live in your selected system. Enter this value into your *Aquaponics Cost and Revenue Analysis* spreadsheet.

Name ______________________________ Date __________ Class Period ______

8. **Fish income:** Show your calculation for the annual income of the fish in your system. You will need to consider how often you can harvest your fish. Enter this value into your *Aquaponics Cost and Revenue Analysis* spreadsheet.

9. **What is your estimated annual revenue for the first year of business?**

10. **How would your annual revenue change in the second year, and why?**

11. **What costs and sources of income are we ignoring?**

LESSON 10
Building an Aquaponics Business

Common Plants and Fish in Aquaponics

Name ______________________________ **Date** __________ **Class Period** ______

Plant	*Optimal Temperature (°F)*[1]	*Optimal pH*[1]	*Extra nutrients?*	*Difficult?*	*Other Information*
Basil	68–77	5.5–6.5	No	No	Often a high-demand plant
Cilantro	50–75	6.5–6.7	No	No	"Bolts" when the temperature is too warm
Lettuce	70–74	5.8–6.2 (7.0)	No	No	Can grow outside listed pH range; can grow in tight spaces; might show iron deficiency at high pH
Spinach	55–75	6.0–7.0	No	No	Easy to care for
Kale	55–70	6.0–7.5	No	No	Potential for aphids; requires a lot of space
Tomato	75–85	5.5–6.5	Yes	Yes	Attracts a lot of pests; tall plants; needs extra potassium for fruiting
Cucumber	72–82	5.5–6.5	Yes	Yes	Good for DWC systems; likes high humidity; needs extra potassium for fruiting
Bell pepper	60–75	5.5–6.5 (7.0)	High nitrate	Yes	Not good in DWC systems; likes warmer air; can tolerate pH of 7
Strawberry	60–80	5.5–6.5	Yes	Yes	Not great for DWC systems; often need additional potassium and iron

Fish	*Optimal Temperature (°F)*[1]	*Optimal pH*[1]	*Extra nutrients?*	*Difficult?*	*Other Information*
Tilapia	65–85	6.5–9.0	Yes	Yes	Easy to care for and popular fish; fairly easy to breed
Koi	50–75	7.0–8.0	Yes	No	Beautiful fish with long life; can sell adults as ornamental fish
Trout	45–65	6.5–8.0	Yes	Yes	Needs lots of space and high DO levels; can't keep with other fish
Bluegill	70–75	7.0–9.0	Yes	Yes	Requires little maintenance
Goldfish	78–82	6.0–8.0	Yes	No	Easy to care for
Shrimp	75–85	7.5–8.5	No	Yes	Susceptible to disease and temperature changes; can be cannibals; will eat waste, algae, or dead roots; good for mature systems

1. The temperature and pH ranges listed are for optimal growth and health. Many of these species can survive outside of the listed ranges; however their growth and health may not be optimized.

LESSON 10

Building an Aquaponics Business

Aquaponics Cost and Revenue Analysis

Name ______________________________ **Date** __________ **Class Period** ______

Based on your selected system, calculate your start-up, testing, and energy costs. Put totals in these boxes.

System Cost	
Water Cost	
Energy Cost	

System	*Cost ($)*	*Tank Size (gallons)*	*No. of Plants*	*Water Tests (annual $ cost)*	*Annual Energy for Lights (kWh[1])*	*Annual Energy for Heat (kWh[1])*	*Energy Cost ($/kWh[1])*
Small aquaponics system	8,000	200	300	2,000	6,000	2,400	0.08
Medium aquaponics system	15,000	600	700	2,000	14,000	5,600	0.08
Large aquaponics system	20,000	1,000	1,500	2,000	30,000	12,000	0.08

1. kWh stands for kilowatt hour and is a common unit for energy/electrical costs.

Name ______________________________ Date __________ Class Period ______

Based on your selected plants, calculate the annual cost and annual income. Put totals in these boxes.

Plant Cost	
Plant Income	

Plant	*Seed Cost ($/seed)*	*Growing Time (days)*	*Selling Price ($/plant)*
Basil	0.02	50	7.00
Cilantro	0.02	50	5.00
Lettuce	0.01	28	2.00
Spinach	0.03	40	3.00
Kale	0.03	42	3.00
Tomato	0.02	60	7.00
Cucumber	0.02	65	6.00
Bell pepper	0.03	80	6.00

Name ______________________________ Date __________ Class Period ______

Based on your selected fish, calculate the annual cost and annual income. Put totals in these boxes.

Fish Cost	
Fish Income	

Fish	*Fish Cost ($/fingerling)*	*Feed Cost ($/ fish per year)*	*Space Requirement (gallons per adult fish)*	*Growing Time (days to harvest)*	*Selling Price ($/fish)*
Tilapia	0.70	7.00	4	220	$2.50
Koi	2.00	10.00	100[1]	N/A[2]	N/A[3]
Trout	1.50	13.00	40	390	$9.00
Bluegill	1.00	8.00	11	365	$6.00
Goldfish	0.30	1.50	6	N/A	N/A
Shrimp	0.50	N/A	0.5	120	$1.50

1. This is for a small koi; large koi require 500 gallons per fish.
2. Koi can live for 20–30 years.
3. If interested in raising koi, research selling prices for adult koi.

Item	*Start-Up Cost (capital expense)*	*Item*	*Annual Operating Cost*	*Item*	*Annual Income*	
System cost		Water tests		Plants		
		Energy		Fish		
		Plants				
		Fish				Annual Revenue
Totals						

LESSON 10

Building an Aquaponics Business

Worksheet Answer Key

Aquaponics Business Challenge

Name ______________________________ **Date** __________ **Class Period** ______

Business Name: ________________________________

1. **Proposal:** Imagine you are starting an aquaponics business, and you require investors to get your business started. Use the space below to write a proposal to persuade them to invest in your company. Start by determining the purpose and goals of your aquaponics system. Is it for profit? Is it for educational purposes? What community will it serve, and why would it be important for that community? Where will it be built? How will the size of the system as well as the fish and plants in your system meet these goals? Your selected fish and plants must be compatible; include the information necessary to confirm your system is viable (see the spreadsheet *Common Plants and Fish in Aquaponics*). Include the projected revenue for your system. If it is negative, justify why your system should be subsidized by the community.

   A number of answers are possible.

Name ______________________________ Date __________ Class Period ______

2. **Economic plan:** Using the information provided in the two spreadsheets (*Common Plants and Fish in Aquaponics* and *Aquaponics Cost and Revenue Analysis*), use the sections below to calculate the start-up costs, annual operating costs, and annual income for your proposed business. If you have decided to grow plants or raise fish that are not on these spreadsheets, complete the research necessary to calculate their cost and income.

   A number of answers are possible.

3. **Start-up cost:** What size system did you select?

   A number of answers are possible.

4. **Energy cost:** Show your calculation for the annual energy cost for your selected system. Enter this value into your spreadsheet.

   A number of answers are possible.

5. **Plant cost:** Show your calculation for the annual cost of plants in your system. Enter this value into your spreadsheet.

   A number of answers are possible.

6. **Plant income:** Show your calculation for the annual income from the plants in your system. Make sure to account for the growing time and number of harvests. Enter this value into your spreadsheet.

   A number of answers are possible.

7. **Fish cost:** Show your calculation for the annual cost of the fish in your system. You will need to calculate how many of your selected fish can live in your selected system. Enter this value into your spreadsheet.

   A number of answers are possible.

8. **Fish income:** Show your calculation for the annual income of the fish in your system. You will need to consider how often you can harvest your fish. Enter this value into your spreadsheet.

   A number of answers are possible.

9. **What is your estimated annual revenue for the first year of business?**

   A number of answers are possible.

10. **How would your annual revenue change in the second year, and why?**

    A number of answers are possible.

11. **What costs and sources of income are we ignoring?**

    A number of answers are possible.

APPENDIX

Standards and Alignment

Lesson 1: Introduction to Aquaponics

 NGSS Alignment[1]

HS-ETS1-3.	Evaluate a solution to a complex real-world problem based on prioritized criteria and trade-offs that account for a range of constraints, including cost, safety, reliability, and aesthetics as well as possible social, cultural, and environmental impacts.

Additional NGSS Alignment

SEP: Science & Engineering Practices	**DCI:** Disciplinary Core Ideas	**CCC:** Crosscutting Concepts
Developing and Using Models Develop, revise, and/or use a model based on evidence to illustrate and/or predict the relationships between systems or between components of a system. **Constructing Explanations and Designing Solutions** Apply scientific ideas, principles, and/or evidence to provide an explanation of phenomena and solve design problems, taking into account possible unanticipated effects.	**PS3.A: Definitions of Energy** Energy is a quantitative property of a system that depends on the motion and interactions of matter and radiation within that system. That there is a single quantity called energy is due to the fact that a system's total energy is conserved, even as, within the system, energy is continually transferred from one object to another and between its various possible forms. **PS3.B: Conservation of Energy and Energy Transfer** Conservation of energy means that the total change of energy in any system is always equal to the total energy transferred into or out of the system. Energy cannot be created or destroyed, but it can be transported from one place to another and transferred between systems.	**Systems and System Models** Models (e.g., physical, mathematical, computer models) can be used to simulate systems and interactions—including energy, matter, and information flows—within and between systems at different scales. **Systems and System Models** When investigating or describing a system, the boundaries and initial conditions of the system need to be defined and their inputs and outputs analyzed and described using models. **Cause and Effect** Cause and effect relationships can be suggested and predicted for complex natural and human-designed systems by examining what is known about smaller-scale mechanisms within the system.

 Great Lakes Literacy Principles[2]

Principle 04: Water makes Earth habitable; fresh water sustains life on land.

1. https://www.nextgenscience.org/pe/hs-ls2-4-ecosystems-interactions-energy-and-dynamics
2. https://www.cgll.org/principles/

Lesson 2: Anatomy of an Aquaponics System

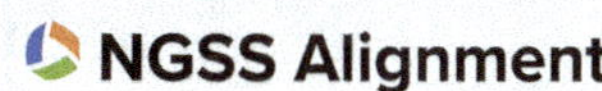

NGSS Alignment[1]

HS-LS2-4.	Use mathematical representations to support claims for the cycling of matter and flow of energy among organisms in an ecosystem.

Additional NGSS Alignment

SEP: Science & Engineering Practices	**DCI:** Disciplinary Core Ideas	**CCC:** Crosscutting Concepts
Developing and Using Models Develop, revise, and/or use a model based on evidence to illustrate and/or predict the relationships between systems or between components of a system. **Constructing Explanations and Designing Solutions** Apply scientific ideas, principles, and/or evidence to provide an explanation of phenomena and solve design problems, taking into account possible unanticipated effects.	**PS3.A: Definitions of Energy** Energy is a quantitative property of a system that depends on the motion and interactions of matter and radiation within that system. That there is a single quantity called energy is due to the fact that a system's total energy is conserved, even as, within the system, energy is continually transferred from one object to another and between its various possible forms. **PS3.B: Conservation of Energy and Energy Transfer** Conservation of energy means that the total change of energy in any system is always equal to the total energy transferred into or out of the system. Energy cannot be created or destroyed, but it can be transported from one place to another and transferred between systems.	**Systems and System Models** Models (e.g., physical, mathematical, computer models) can be used to simulate systems and interactions—including energy, matter, and information flows—within and between systems at different scales. **Systems and System Models** When investigating or describing a system, the boundaries and initial conditions of the system need to be defined and their inputs and outputs analyzed and described using models. **Cause and Effect** Cause and effect relationships can be suggested and predicted for complex natural and human-designed systems by examining what is known about smaller-scale mechanisms within the system.

Great Lakes Literacy Principles[2]

Principle 04: Water makes Earth habitable; fresh water sustains life on land.

1. https://www.nextgenscience.org/pe/hs-ls2-4-ecosystems-interactions-energy-and-dynamics
2. https://www.cgll.org/principles/

Lesson 03: Introduction to the Nitrogen Cycle

NGSS Alignment[1]

HS-LS-2-4.	Use mathematical representations to support claims for the cycling of matter and flow of energy among organisms in an ecosystem.

Additional NGSS Alignment

SEP: Science & Engineering Practices	DCI: Disciplinary Core Ideas	CCC: Crosscutting Concepts
Developing and Using Models Develop, revise, and/or use a model based on evidence to illustrate and/or predict the relationships between systems or between components of a system. **Constructing Explanations and Designing Solutions** Apply scientific ideas, principles, and/or evidence to provide an explanation of phenomena and solve design problems, taking into account possible unanticipated effects. **Using Mathematics and Computational Thinking** Use mathematical, computational, and/or algorithmic representations of phenomena or design solutions to describe and/or support claims and/or explanations **Engaging in Argument from Evidence** Construct, use, and/or present an oral and written argument or counter-arguments based on data and evidence.	**PS3.A: Definitions of Energy** Energy is a quantitative property of a system that depends on the motion and interactions of matter and radiation within that system. That there is a single quantity called energy is due to the fact that a system's total energy is conserved, even as, within the system, energy is continually transferred from one object to another and between its various possible forms. **PS3.B: Conservation of Energy and Energy Transfer** Conservation of energy means that the total change of energy in any system is always equal to the total energy transferred into or out of the system. Energy cannot be created or destroyed, but it can be transported from one place to another and transferred between systems.	**Patterns** Different patterns may be observed at each of the scales at which a system is studied and can provide evidence for causality in explanations of phenomena. **Cause and Effect** Cause and effect relationships can be suggested and predicted for complex natural and human designed systems by examining what is known about smaller scale mechanisms within the system. **Systems and System Models** Models (e.g., physical, mathematical, computer models) can be used to simulate systems and interactions—including energy, matter, and information flows—within and between systems at different scales.

Great Lakes Literacy Principles[2]

Principle 04: Water makes Earth habitable; fresh water sustains life on land.

1. https://www.nextgenscience.org/pe/hs-ls2-4-ecosystems-interactions-energy-and-dynamics
2. https://www.cgll.org/principles/

Lesson 04: Measuring Nitrogen Levels in Your System

NGSS Alignment[1]

HS-LS-2-4.	Use mathematical representations to support claims for the cycling of matter and flow of energy among organisms in an ecosystem.

Additional NGSS Alignment

SEP: Science & Engineering Practices	**DCI:** Disciplinary Core Ideas	**CCC:** Crosscutting Concepts
Using Mathematics and Computational Thinking Apply techniques of algebra and functions to represent and solve scientific and engineering problems. **Planning and Conducting Investigations** Plan an investigation or test a design individually and collaboratively to produce data to serve as the basis for evidence as part of building and revising models, supporting explanations for phenomena, or testing solutions to problems. Consider possible confounding variables or effects and evaluate the investigation's design to ensure variables are controlled.	**PS3.A: Definitions of Energy** Energy is a quantitative property of a system that depends on the motion and interactions of matter and radiation within that system. That there is a single quantity called energy is due to the fact that a system's total energy is conserved, even as, within the system, energy is continually transferred from one object to another and between its various possible forms. **PS3.B: Conservation of Energy and Energy Transfer** Conservation of energy means that the total change of energy in any system is always equal to the total energy transferred into or out of the system. Energy cannot be created or destroyed, but it can be transported from one place to another and transferred between systems.	**Scale, Proportion and Quantity** Algebraic thinking is used to examine scientific data and predict the effect of a change in one variable on another (e.g., linear growth vs. exponential growth). **Stability and Change** Much of science deals with constructing explanations of how things change and how they remain stable.

Great Lakes Literacy Principles[2]

Principle 04: Water makes Earth habitable; fresh water sustains life on land.

1. https://www.nextgenscience.org/pe/hs-ls2-4-ecosystems-interactions-energy-and-dynamics
2. https://www.cgll.org/principles/

Lesson 05: Nitrogen Cycle & Population Dynamics

NGSS Alignment[1]

HS-LS-2-2.	Use mathematical representations to support and revise explanations based on evidence about factors affecting biodiversity and populations in ecosystems of different scales.

Additional NGSS Alignment

SEP: Science & Engineering Practices	**DCI:** Disciplinary Core Ideas	**CCC:** Crosscutting Concepts
Analyzing and Interpreting Data Analyze data to identify design features or characteristics of the components of a proposed process or system to optimize it relative to criteria for success. **Asking Questions and Defining Problems** Ask questions to determine relationships, including quantitative relationships, between independent and dependent variables **Constructing Explanations and Designing Solutions** Apply scientific reasoning, theory, and/or models to link evidence to the claims to assess the extent to which the reasoning and data support the explanation or conclusion.	**PS3.A: Definitions of Energy** Energy is a quantitative property of a system that depends on the motion and interactions of matter and radiation within that system. That there is a single quantity called energy is due to the fact that a system's total energy is conserved, even as, within the system, energy is continually transferred from one object to another and between its various possible forms. **PS3.B: Conservation of Energy and Energy Transfer** Conservation of energy means that the total change of energy in any system is always equal to the total energy transferred into or out of the system. Energy cannot be created or destroyed, but it can be transported from one place to another and transferred between systems.	**Patterns** Different patterns may be observed at each of the scales at which a system is studied and can provide evidence for causality in explanations of phenomena. **Cause and Effect** Cause and effect relationships can be suggested and predicted for complex natural and human designed systems by examining what is known about smaller scale mechanisms within the system.

Great Lakes Literacy Principles[2]

Principle 04: Water makes Earth habitable; fresh water sustains life on land.

1. https://www.nextgenscience.org/pe/hs-ls2-2-ecosystems-interactions-energy-and-dynamics
2. https://www.cgll.org/principles/

Lesson 6: Plants in Aquaponic Systems

NGSS Alignment[1]

HS-LS1-5. Use a model to illustrate how photosynthesis transforms light energy into stored chemical energy.

HS-LS2-2. Use mathematical representations to support and revise explanations based on evidence about factors affecting biodiversity and populations in ecosystems of different scales.

HS-LS2-3. Construct and revise an explanation based on evidence for the cycling of matter and flow of energy in aerobic and anaerobic conditions.

HS-LS2-5. Develop a model to illustrate the role of photosynthesis and cellular respiration in the cycling of carbon among the biosphere, atmosphere, hydrosphere, and geosphere.

HS-LS2-4. Use mathematical representations to support claims for the cycling of matter and flow of energy among organisms in an ecosystem.

Additional NGSS Alignment

SEP: Science & Engineering Practices	**DCI:** Disciplinary Core Ideas	**CCC:** Crosscutting Concepts
Analyzing and Interpreting Data Analyze data using tools, technologies, and/or models (e.g., computational, mathematical) in order to make valid and reliable scientific claims or determine an optimal design solution. **Planning and Conducting Investigations** Plan and conduct an investigation individually and collaboratively to produce data to serve as the basis for evidence, and in the design: decide on types, how much, and accuracy of data needed to produce reliable measurements and consider limitations on the precision of the data (e.g., number of trials, cost, risk, time), and refine the design accordingly. **Obtaining, Evaluating and Communicating Information** Communicate scientific and/or technical information or ideas (e.g. about phenomena and/or the process of development and the design and performance of a proposed process or system) in multiple formats (including orally, graphically, textually, and mathematically).	**PS3.A: Definitions of Energy** Energy is a quantitative property of a system that depends on the motion and interactions of matter and radiation within that system. That there is a single quantity called energy is due to the fact that a system's total energy is conserved, even as, within the system, energy is continually transferred from one object to another and between its various possible forms. **PS3.B: Conservation of Energy and Energy Transfer** Conservation of energy means that the total change of energy in any system is always equal to the total energy transferred into or out of the system. Energy cannot be created or destroyed, but it can be transported from one place to another and transferred between systems.	**Patterns** Patterns of performance of designed systems can be analyzed and interpreted to reengineer and improve the system.

Great Lakes Literacy Principles[2]

Principle 04: Water makes Earth habitable; fresh water sustains life on land.

1. https://www.nextgenscience.org/
2. https://www.cgll.org/principles/

Lesson 7: Fish in Aquaponics Systems

NGSS Alignment[1]

HS-ETS1-3.	Evaluate a solution to a complex real-world problem based on prioritized criteria and trade-offs that account for a range of constraints, including cost, safety, reliability, and aesthetics as well as possible social, cultural, and environmental impacts.

Additional NGSS Alignment

SEP: Science & Engineering Practices	DCI: Disciplinary Core Ideas	CCC: Crosscutting Concepts
Constructing Explanations and Designing Solutions Design, evaluate, and/or refine a solution to a complex real-world problem, based on scientific knowledge, student-generated sources of evidence, prioritized criteria, and tradeoff considerations. **Obtaining, Evaluating, and Communicating Information** Compare, integrate and evaluate sources of information presented in different media or formats (e.g., visually, quantitatively) as well as in words in order to address a scientific question or solve a problem. **Modeling** Develop and/or use multiple types of models to provide mechanistic accounts and/or predict phenomena, and move flexibly between model types based on merits and limitations **Analyzing and Interpreting Data** Analyze data using tools, technologies, and/or models (e.g., computational, mathematical) in order to make valid and reliable scientific claims or determine an optimal design solution.	**PS3.A: Definitions of Energy** Energy is a quantitative property of a system that depends on the motion and interactions of matter and radiation within that system. That there is a single quantity called energy is due to the fact that a system's total energy is conserved, even as, within the system, energy is continually transferred from one object to another and between its various possible forms. **PS3.B: Conservation of Energy and Energy Transfer** Conservation of energy means that the total change of energy in any system is always equal to the total energy transferred into or out of the system. Energy cannot be created or destroyed, but it can be transported from one place to another and transferred between systems.	**Patterns** Different patterns may be observed at each of the scales at which a system is studied and can provide evidence for causality in explanations of phenomena. **Cause and Effect** Cause and effect relationships can be suggested and predicted for complex natural and human designed systems by examining what is known about smaller scale mechanisms within the system. **System and System Models** Models (e.g., physical, mathematical, computer models) can be used to simulate systems and interactions—including energy, matter, and information flows—within and between systems at different scales. **Structure and Function** Investigating or designing new systems or structures requires a detailed examination of the properties of different materials, the structures of different components, and connections of components to reveal its function and/or solve a problem.

Great Lakes Literacy Principles[2]

Principle 04: Water makes Earth habitable; fresh water sustains life on land.

1. https://www.nextgenscience.org/pe/hs-ls2-2-ecosystems-interactions-energy-and-dynamics
2. https://www.cgll.org/principles/

Lesson 8: Our Modern Food System

NGSS Alignment[1]

HS-ETS1-3.	Evaluate a solution to a complex real-world problem based on prioritized criteria and trade-offs that account for a range of constraints, including cost, safety, reliability, and aesthetics as well as possible social, cultural, and environmental impacts.

Additional NGSS Alignment

SEP: Science & Engineering Practices	**DCI:** Disciplinary Core Ideas	**CCC:** Crosscutting Concepts
Constructing Explanations and Designing Solutions Design, evaluate, and/or refine a solution to a complex real-world problem, based on scientific knowledge, student-generated sources of evidence, prioritized criteria, and tradeoff considerations. **Obtaining, Evaluating, and Communicating Information** Compare, integrate and evaluate sources of information presented in different media or formats (e.g., visually, quantitatively) as well as in words in order to address a scientific question or solve a problem. **Modeling** Develop and/or use multiple types of models to provide mechanistic accounts and/or predict phenomena, and move flexibly between model types based on merits and limitations	**PS3.A: Definitions of Energy** Energy is a quantitative property of a system that depends on the motion and interactions of matter and radiation within that system. That there is a single quantity called energy is due to the fact that a system's total energy is conserved, even as, within the system, energy is continually transferred from one object to another and between its various possible forms. **PS3.B: Conservation of Energy and Energy Transfer** Conservation of energy means that the total change of energy in any system is always equal to the total energy transferred into or out of the system. Energy cannot be created or destroyed, but it can be transported from one place to another and transferred between systems.	**Cause and Effect** Cause and effect relationships can be suggested and predicted for complex natural and human designed systems by examining what is known about smaller scale mechanisms within the system. **System and System Models** Models (e.g., physical, mathematical, computer models) can be used to simulate systems and interactions—including energy, matter, and information flows—within and between systems at different scales. **Structure and Function** Investigating or designing new systems or structures requires a detailed examination of the properties of different materials, the structures of different components, and connections of components to reveal its function and/or solve a problem.

Great Lakes Literacy Principles[2]

Principle 04: Water makes Earth habitable; fresh water sustains life on land.

1. https://www.nextgenscience.org/pe/hs-ets1-3-engineering-design
2. https://www.cgll.org/principles/

Lesson 09: Troubleshooting Water Quality Challenges

NGSS Alignment[1]

HS-ETS1-3.	Evaluate a solution to a complex real-world problem based on prioritized criteria and trade-offs that account for a range of constraints, including cost, safety, reliability, and aesthetics as well as possible social, cultural, and environmental impacts.

Additional NGSS Alignment

SEP: Science & Engineering Practices	DCI: Disciplinary Core Ideas	CCC: Crosscutting Concepts
Constructing Explanations and Designing Solutions Design, evaluate, and/or refine a solution to a complex real-world problem, based on scientific knowledge, student-generated sources of evidence, prioritized criteria, and tradeoff considerations. **Obtaining, Evaluating, and Communicating Information** Compare, integrate and evaluate sources of information presented in different media or formats (e.g., visually, quantitatively) as well as in words in order to address a scientific question or solve a problem. **Analyzing and Interpreting Data** Analyze data using tools, technologies, and/or models (e.g., computational, mathematical) in order to make valid and reliable scientific claims or determine an optimal design solution. Analyze data to identify design features or characteristics of the components of a proposed process or system to optimize it relative to criteria for success. **Planning and Carrying Out Investigations** Manipulate variables and collect data about a complex model of a proposed process or system to identify failure points or improve performance relative to criteria for success or other variables.	**PS3.A: Definitions of Energy** Energy is a quantitative property of a system that depends on the motion and interactions of matter and radiation within that system. That there is a single quantity called energy is due to the fact that a system's total energy is conserved, even as, within the system, energy is continually transferred from one object to another and between its various possible forms. **PS3.B: Conservation of Energy and Energy Transfer** Conservation of energy means that the total change of energy in any system is always equal to the total energy transferred into or out of the system. Energy cannot be created or destroyed, but it can be transported from one place to another and transferred between systems.	**Patterns** Patterns of performance of designed systems can be analyzed and interpreted to reengineer and improve the system. **Cause and Effect** Cause and effect relationships can be suggested and predicted for complex natural and human designed systems by examining what is known about smaller scale mechanisms within the system. **System and System Models** Models (e.g., physical, mathematical, computer models) can be used to simulate systems and interactions—including energy, matter, and information flows—within and between systems at different scales.

Great Lakes Literacy Principles[2]

Principle 04: Water makes Earth habitable; fresh water sustains life on land.

1. https://www.nextgenscience.org/pe/hs-ets1-3-engineering-design
2. https://www.cgll.org/principles/

Lesson 10: Building an Aquaponics Business

NGSS Alignment[1]

HS-ETS1-3.	Evaluate a solution to a complex real-world problem based on prioritized criteria and trade-offs that account for a range of constraints, including cost, safety, reliability, and aesthetics as well as possible social, cultural, and environmental impacts.

Additional NGSS Alignment

SEP: Science & Engineering Practices	DCI: Disciplinary Core Ideas	CCC: Crosscutting Concepts
Asking Questions and Defining Problems Define a design problem that involves the development of a process or system with interacting components and criteria and constraints that may include social, technical and/or environmental considerations. **Analyzing and Interpreting Data** Develop a complex model that allows for manipulation and testing of a proposed process or system. **Developing and Using Models** Develop a complex model that allows for manipulation and testing of a proposed process or system. **Using Mathematics and Computational Thinking** Use mathematical, computational, and/or algorithmic representations of phenomena or design solutions to describe and/or support claims and/or explanations **Engaging in Argument from Evidence** Evaluate competing design solutions to a real-world problem based on scientific ideas and principles, empirical evidence, and/or logical arguments regarding relevant factors (e.g. economic, societal, environmental, ethical considerations).	**PS3.A: Definitions of Energy** Energy is a quantitative property of a system that depends on the motion and interactions of matter and radiation within that system. That there is a single quantity called energy is due to the fact that a system's total energy is conserved, even as, within the system, energy is continually transferred from one object to another and between its various possible forms. **PS3.B: Conservation of Energy and Energy Transfer** Conservation of energy means that the total change of energy in any system is always equal to the total energy transferred into or out of the system. Energy cannot be created or destroyed, but it can be transported from one place to another and transferred between systems.	**Patterns** Patterns of performance of designed systems can be analyzed and interpreted to reengineer and improve the system. **Cause and Effect** Systems can be designed to cause a desired effect. Systems and System Models Models (e.g., physical, mathematical, computer models) can be used to simulate systems and interactions—including energy, matter, and information flows—within and between systems at different scales. **Energy and Matter** Energy drives the cycling of matter within and between systems.

Great Lakes Literacy Principles[2]

Principle 04: Water makes Earth habitable; fresh water sustains life on land.

1. https://www.nextgenscience.org/pe/hs-ets1-3-engineering-design
2. https://www.cgll.org/principles/

About the Authors

AMY M. SHAMBACH is an Extension associate at Purdue University with the Illinois-Indiana Sea Grant (IISG) Program. She has a background in aquaculture production and works with the aquaculture industry in the USDA's North Central Region. Her current work focuses on the demand side of domestic farm-raised seafood products, consumer education, aquaculture production, and K–12 education.

REBECCA M. WARD is a lecturer in general chemistry at The Ohio State University. Prior to joining OSU, Ward built aquaponic systems in K–12 public schools, designed accompanying curriculum, wrote scripts for educational science videos, and oversaw the upper-division chemistry labs at Mississippi State University.

J. STUART CARLTON is the director of the Illinois-Indiana Sea Grant (IISG) College Program and head of the Coastal and Great Lakes Social Science Lab in the Department of Forestry and Natural Resources at Purdue University, where he also serves as a research assistant professor. His lab group studies the relationship between knowledge, values, trust, and behavior in complex or controversial environmental systems.

KWAMENA K. QUAGRAINIE is a professor of aquaculture economics and an aquaculture marketing specialist at Purdue University. He is also affiliated with the Illinois-Indiana Sea Grant (IISG) College Program. His research is interdisciplinary and focuses on market analysis, market definition, identification of value-added opportunities for aquaculture products, and the economic viability of aquaculture and aquaponics farming enterprises.

TOMAS O. HÖÖK is professor and department head in the Department of Forestry and Natural Resources at Purdue University, where his research and teaching areas of expertise include fisheries and aquatic sciences of the Laurentian Great Lakes. Prior to joining Purdue, he worked at the Cooperative Institute for Limnology and Ecosystems Research. Höök previously served as the associate director of research and then director of Illinois-Indiana Sea Grant.

www.ingramcontent.com/pod-product-compliance
Lightning Source LLC
LaVergne TN
LVHW061246100826
845148LV00008B/1045

* 9 7 8 1 6 2 6 7 1 2 4 7 8 *